SQL Server 2014 Backup and Recovery

Techniques for Backing up and Restoring Databases in SQL Server 2014

By

Tim Radney & John Sterrett

ISBN: 978-1502573896

COVER PHOTO BY TODD RADNEY OF YESTERDAYS PHOTOGRAPHY

About the Authors

Tim Radney

Tim Radney is a Microsoft "Most Valuable Professional" MVP, which has been working in IT for nearly two decades. He has spent the past 6 years focusing on Microsoft SQL Server and currently holds certifications across many aspects of IT. He is dedicated to paying it forward within the SQL Server community. He serves as the chapter leader for the Columbus, GA SQL Users Group, is a PASS Regional Mentor for the Greater Southeast USA and is a regular speaker at SQL user groups, SQL Saturdays and the PASS Summit.

Tim blogs at http://www.timradney.com and tweets at https://www.twitter.com/tradney

John Sterrett

John Sterrett is an independent consultant focusing on solving business problems with SQL Server. Previously, he was a Sr. Database Admin Advisor for Dell, and was a Senior Database Admin for Remote DBA Experts. He serves as the chapter leader for the Austin, TX SQL User Group and the High Availability and Disaster Recovery Virtual Chapter. He is a PASS Regional Mentor and a regular speaker at SQL user groups, SQL Saturdays and IT Conferences like the PASS Member Summit.

Acknowledgements from Tim Radney

First I have to give recognition to my wife Theresa, who is my biggest fan. She inspires me like no one ever has. Our journey together began when we were put on the same project at work. We became very good friends and our relationship grew from there. We have three amazing kids, Gracie, Jordan and Riley that keep us on the go.

I would like to thank my parents Scott and Caffey Radney for all their love and support. I wouldn't be the person I am today without you both inspiring me.

A huge thank you to my brother Todd Radney of Yesterdays Photography for taking the awesome cover photo and allowing me to use it for my book.

A big thanks to the entire SQL Server community. Without you all, I would be completely lost in this journey. When I started becoming a fulltime DBA, it was my SQL Server family who mentored me and helped me become what I am now. Not only has my SQL Server family helped with me the technical aspects of SQL Server, but my mentors have taught me to be a better manager, dad, husband, and chicken farmer. Thank you to my entire Linchpin family for your daily support. I couldn't have accomplished any of this without them. #sqlfamily

To Rick Morelan for enabling me to write my first book.

To Karen Forster for being an amazing editor who helped make this book happen. Had she been editing with pen, I would owe her a lot of red ink.

Acknowledgements from John Sterrett

To my wife Nina, thank you for your love, support and patience.

To my son Gregory, thank you for always finding a way to make me smile and laugh.

To my parents, thank you for everything. Denise, thank you for giving me my first SQL Server book. Cedric, thank you for always listening to anything I say.

To my SQL Family and the Greater Wheeling Chapter of AITP, thank you giving me the opportunity to connect, share and learn with you.

To Brian and Andy, thank you for believing in me and mentoring me to be a better SQL person.

To Tim Radney, thank you for giving me an opportunity to work on this book with you.

To Karen Forster, thank you for excellent advice and guidance with editing my work.

About the Editor

Karen Forster has more than 20 years' experience as an IT industry journalist and technology content expert. Most recently, Karen worked at Microsoft as Director of Technical Communications and earlier as Director of Windows Server documentation. Previously, Karen was Editorial Director of Windows IT Pro magazine and SQL Server Magazine, as well as numerous online publications and web sites. Karen's goal is to serve the IT industry and community by providing outstanding independent content and perspective.

About this Book

This book is designed to help the beginner and mid-level DBA to get a strong understanding of the types of backups available within SQL Server and how to restore each of those backup types.

This book will cover each type of the backup and then go over several common restore scenarios. The related examples will give readers practical experience in performing the different backup and restore techniques.

Working in the IT industry thrusts you into stressful situations constantly. You learn many skills on the job during these stressful situations. The one skill that you do not want to learn in a crisis is how to recover your database. Being able to restore and bring a database online is basic knowledge for DBAs. The goal of this book is that after reading it and following along with the exercises, you will be comfortable with restoring any type of backup.

Skills Needed for this Book

This book is designed as a focused specialized lesson in good practices for backing up a database. Most readers of this book should be familiar with SQL Server and its general functions, as well as how to navigate the object explorer. Readers should have at least have basic Transact-SQL (T-SQL) skills. Many readers may also have done many types of backups but want to know more about all types of backups and when and how to use them together. You can be a beginner at either SQL Server administration or SQL Server development, as long as you are at an intermediate level in the other one. So welcome SQL Server admins and devs to honing your backup skills.

Software Needed for this Book

You will need to have a system with SQL Server 2014 installed. We recommend SQL Server Developer Edition as the lowest cost way to be able to do every example in this book. You can find SQL Server Developer Edition 2014 on Amazon for around $60. If you use SQL Server Express, then about 20 percent of the labs and examples will not work for you.

Chapter 1. SQL Server Database Files

Being a database administrator (DBA) can be a very rewarding career. It can also become a very short career if you don't master recovery and backups. Notice that I listed recovery first, even though you can't recover if you don't have a backup. Backups are the foundation of a disaster recovery plan, but they don't provide any value to your users or your career if you can't restore the backups correctly to recover with as little data loss as possible.

My introduction to database restores convinced me of the importance that backup and recovery have in a DBA's career. In my career, I started out as a developer. I was always fascinated with the code to access data and with making that code as fast as possible. During my introduction to the database administration world, the company I worked for had a storage problem in the data center, and I got the job of recovering from backups.

At this point in my career, I had never done a database restore for a corrupt database. Several people were looking over my shoulders, and I was sweating bullets.

I will never forget watching the status of the restore and wondering how long this restore would take. Would it actually complete successfully? Whenever the progress of the restore wasn't moving, I was starting to wonder if I should be updating my resume. Lucky for me, I was eventually able to recover and bring the critical database back online.

This experience made me realize that I needed to put all my other database administrator objectives on hold until I fully understood backups and restores because these skills would be the lifeline for a long successful career as a database administrator. This book is Tim's and my effort to help our fellow DBAs master backup and recovery skills.

This chapter will start by discussing the basics of SQL Server database files. Then it will dive into the various types of backups: file, full, differential, copy, and transaction log backups.

SQL Server Databases and Database Files: The Basics

SQL Server databases have two different types of files: data files and transactional log files. Data files typically have an MDF or NDF file extension. Transactional log files typically have LDF file extension. (However, there is no formal requirement for data files ending with these extensions.) When you create a new database, you have an empty database file and transactional log file. It contains no data. It doesn't include any user tables, either.

When you create tables and implement a transaction to insert rows, SQL Server records these transactions in the transaction log and then uses 8KB blocks, known as pages, to store the data into a database file. Eight database pages, or 64KB, are then grouped together into an extent. The transactional log records all transactions and the database modifications made with each transaction.

For example, say you had a table called *leads* and you ran the following statement:

```
SELECT * FROM leads
```

You could get a current snapshot, and see all the rows in leads. What you wouldn't see is that a specific lead named "Linchpin People" was created on Monday at 8:00 AM. In fact, you might not see this lead at all in your query because on Tuesday this lead became a customer. Therefore, it was deleted from the leads table and placed in the clients table. The snapshot shown by the query was executed after the lead named "Linchpin People" was moved. This example demonstrates how the transactional log records all transactions and the database modifications made by each transaction.

Figure 1.1 shows the default file layout for the LinchpinPress database you will create and use for the examples in this chapter.

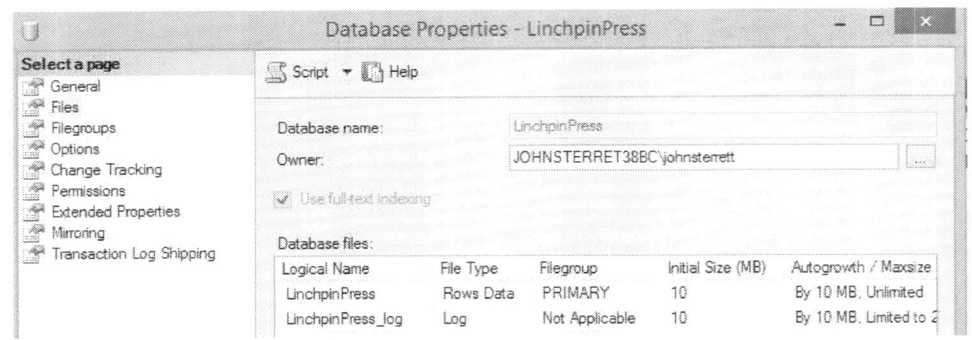

Figure 1.1 Default database and transactional log file created with the LinchpinPress database

File Backups

The first type of SQL Server backup is the file backup, or cold backup. SQL Server file backups are when you copy data files and transactional log files with the SQL Server instance offline. This means that the SQL Server instance is not available to users during a copy backup. Therefore, a file backup is a good way to back up and archive existing SQL Server backup files but not the actual database files since the SQL Server instance must be offline during a file backup.

Figure 1.2 shows the error you get if you try to back up a file that is in use.

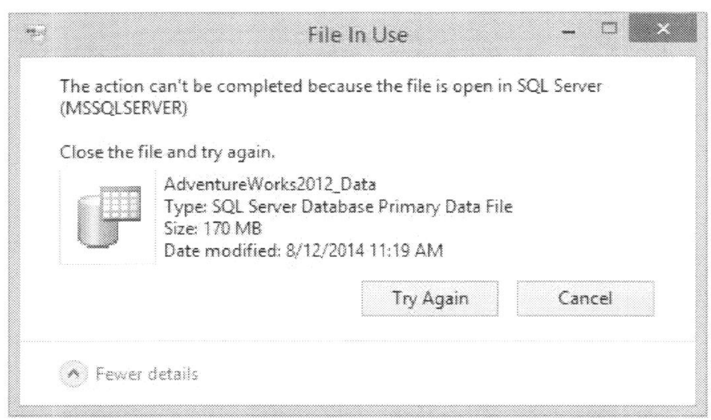

Figure 1.2 You cannot copy a database file that is online while the SQL Server service is running

You don't want to leverage file backups to back up your SQL Server databases, but file backups are a great way to back up your SQL Server backups so they can be archived in case they are needed to be restored.

Full Backups

In SQL Server, a full backup is a copy of the whole database as it existed at the time the backup was completed. Full backups are the basis for other types of backups and are key to recovery planning. You use full backups when you need to back up the whole database, and full backups form the base for differential or transactional log backups.

Let's walk through an example of doing a full backup. This example will have five total extents. (Again, an extent is a collection of eight database pages.) Each row in this example database is its own extent. SQL Server uses a differential bitmap page, which contains a bit for every extent to signal changes to that extent. If an extent has been updated since a full backup established the base for future backups, the *differential bitmap* for the updated extent is set to 1—that is, the extent "raises its flag" to signal that it has been changed. A full backup will back up the entire database, whether or not any flags are raised. Once the full backup completes, the full backup will reset all the flags. (These flags, or differential bitmaps, will become important with other types of backups, which we'll discuss below.)

For this example, assume you have daily full backups, as illustrated in Figure 1.3.

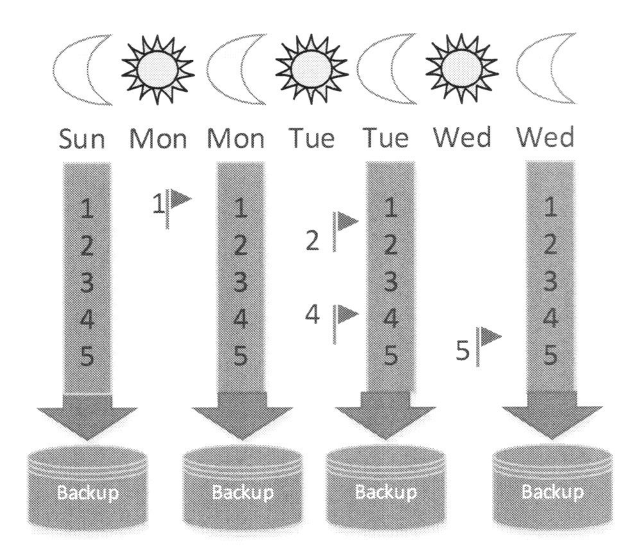

Figure 1.3 Full backups done each night with flags indicating extents with changes from the work day

This example will look at Sunday through Wednesday. A full backup runs every night and will back up the entire set of files Monday evening. During the day, users modify the data. On Monday, users make changes to data in extent 1. The differential bitmap for extent 1 is set to 1, showing extent 1 has been changed. In other words, extent 1 raises its flag to signal that it needs to be backed up on Monday night. Once the backup is done, the flag is taken down, indicating extent 1 has been backed up.

Storage Space and Full Backups

Full backups are a requirement for a backup and recovery strategy. However, a concern that people commonly overlook is the storage space required for full database backups. Not only will you need to be able to recover the backup, but you need to recover as quickly as possible and using the smallest possible amount of files and disk storage. If you perform a full backup daily, you will back up all database pages regardless of how much data has changed since your previous backup.

When you have multiple databases for multiple database servers, backing up all data instead of just data that has changed can require extra storage space that isn't required if you use differential and transactional log backups with the full backup. Let's look at an example to show what this means.

Assume a crash happens Tuesday after the midnight backup but before extents 2 and 4 are modified. Since a full backup occurred on Tuesday at midnight, the database could be restored because no changes would have occurred yet to the data that was backed up Tuesday. If you're doing full backups every night and no data has changed since the last backup, you only need to restore the most recent full backup to recover all the extents.

Keep in mind that this is not a realistic situation. You most likely will have transactions between your full backup and a system crash. When you restore the previous day's full backup, those transactions will cause you to lose data because the changes would not have been captured in the full backup, as Figure 1.4 shows.

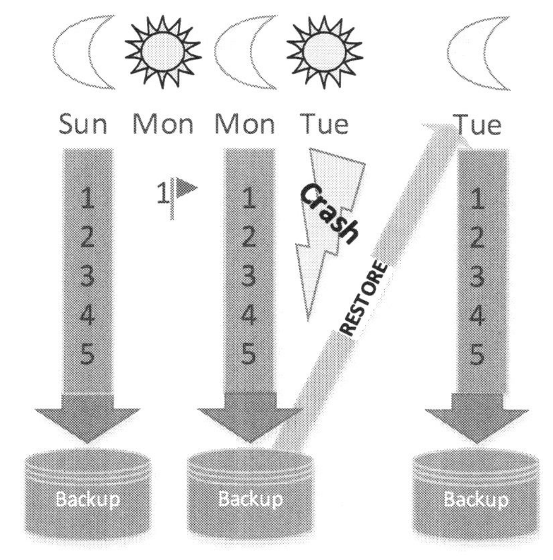

Figure 1.4 The restore process with full backups

Characteristics of Full Backups

Full backups have the following features: All extents are backed up completely, all flags are reset, and the entire database is backed up whether extents were changed or not changed. You use full backups to back up the whole database or to create a base for differential and transactional log backups.

Differential Backup

If you're familiar with systems *other than* SQL Server, you're probably familiar with incremental backups. An incremental backup captures only changes from the existing version of the object being backed up.

For example, if I had thousands of photos and my backup process was as simple as creating a folder with the day's date and copying my photos to that folder, I would not want to copy every single photo every day. Incremental backups would be ideal. I would have a base file that contained all my photos, and then I could back up only new photos that I added on top of that initial set of photos. This approach would save me from having to back up all the existing photos along with all the new photos.

If you like the concept of incremental backups, I have some good news and some bad news. The bad news is that incremental backups do not exist in SQL Server. The good news is that the concept of incremental backups is not too different from SQL Server's differential backups and transactional log backups. (Note for now that transactional log backups and differential backups are very different. We will go over transactional log backups separately in this chapter.)

Differential backups capture the state of any extents that have changed between the time when the differential base (or full backup) was created and when the differential backup is created. In SQL Server, your latest full backup is the base for differential backups. Your latest full backup and the latest corresponding differential backup are the base for transactional log backups.

You use SQL Server differential backups when you want a full backup but don't want the storage overhead of backing up data that hasn't changed since your last full backup. Each differential backup includes data that has changed since the last backup until those changes are captured in the next full backup.

Recall that an extent is a collection of eight database pages. Each extent that has had a change raises its flag, or differential bitmap. A differential backup relies on the differential bitmap page, which contains a bit for every extent. If an extent has been updated since its base was established, the bit for that extent is set to 1 in the bitmap, which means it is marked to be included in the differential backup. This is referred to as the extent raising its flag.

Figure 1.5 illustrates how a differential backup contrasts with a full backup. It shows that instead of doing daily full backups, you can do a weekly full backup on Sunday and a differential backup on each other day.

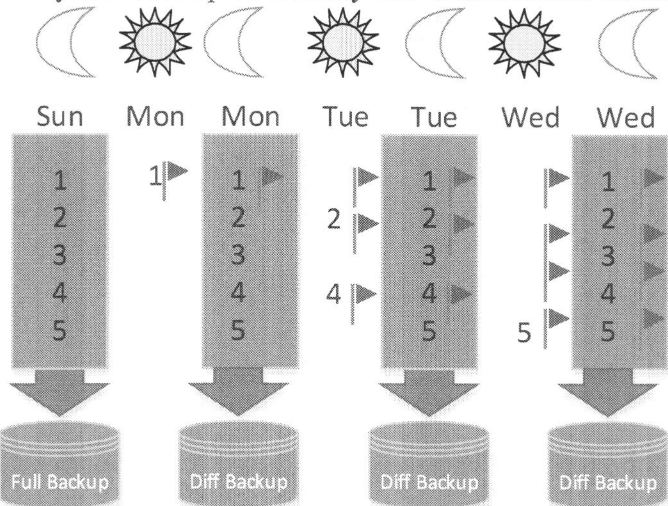

Figure 1.5 Persisting flags after successful differential backups backup

During the day on Monday, extent 1 is modified. Extent 1 had its differential bitmap set; in other words, its flag was raised, indicating that it needs to be included in the differential backup on Monday night. After the differential backup occurs, the differential bitmap is *not* cleared. This is a major contrast between a full backup and a differential backup.

During the day Tuesday, extents 2 and 4 are modified. The differential bitmap is now set for extents 1, 2, and 4. Remember the differential backup does not clear differential bitmaps for extents.

Tuesday night, both extent 2 and extent 4 are backed up. Note that although extent 1 was not changed on Tuesday, extent 1 will also be backed up since its differential bitmap was not cleared at the previous differential backup, as is the normal behavior of a differential backup. Once the backup for Tuesday is successful, the differential bitmaps remain set because they are not cleared during differential backups.

On Wednesday, extent 5 changes. Now Wednesday's backup will include extents 1, 2, 4, and 5. Each day, the differential backup set gets bigger and bigger until the differential bitmaps are cleared by a full backup. In this scenario differential backups continue to grow until the full backup occurs on Sunday, as shown in Figure 1.5.

Suppose that on Thursday morning, the database enters an unhealthy state. A full backup from Sunday and the differential backup from Wednesday are required to restore the database. Sunday's full backup has all the changes from the point in time when the backup was created on Sunday. Wednesday has all the altered extents for the week, as shown in Figure 1.6.

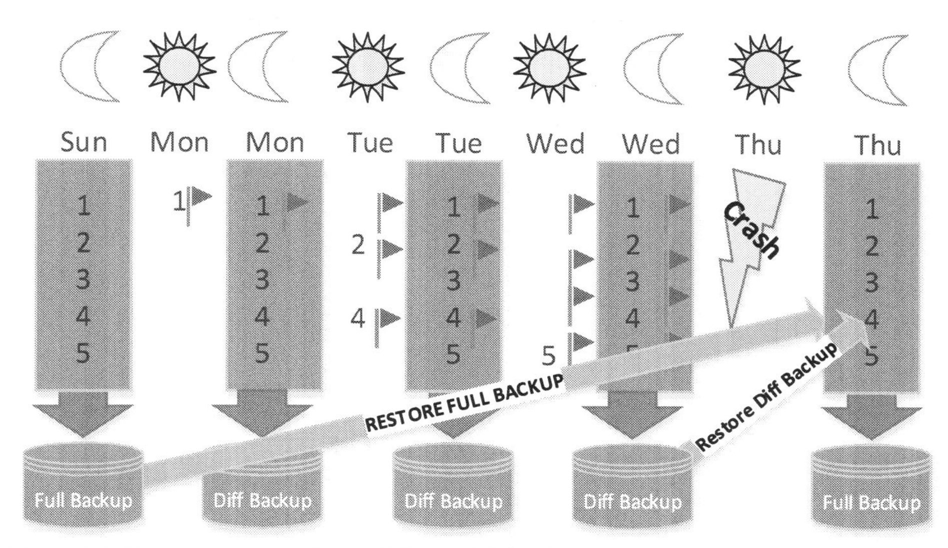

Figure 1.6 Restore process from differential backups

Characteristics of Differential Backups

Differential backups have the following characteristics:

- All extents that have been modified are backed up during each differential backup
- The flags (or differential bitmaps) are not reset following a successful differential backup
- Only two backup file sets (the full backup file and the most recent full and differential backup files) are required

Differential backups get bigger with each backup because all changes that have occurred since the last full backup are backed up. For this reason, it can be confusing when people refer to differential backups as incremental backups, because incremental backups include only changes that have been made since the last backup.

Copy Backup

A copy backup is when you copy the entire database without changing any of the flags. The copy backup is especially useful in a system test because it allows a full restore without interfering with scheduled differential backups. The copy backup can be maintained with as little as one backup file.

What is the potential danger of taking a full backup in the middle of the backup cycle? Remember the differential example: A full database backup occurs on Sundays, and on each other day, a differential backup is done. The differential backup includes all changes for each day, so each day the differential backup grows. When you take a full backup on Wednesday, the differential backup for Thursday will be a lot smaller than Wednesday's backup because the full backup captures all the changes from the previous differential backup and resets all the flags.

Suppose you did the Thursday backup, but not everyone knew about the backup. If another DBA had to do a restore, he or she would restore the full backup from Sunday and try to restore the latest differential backup. It would fail. The error that the uninformed DBA receives is shown in Figure 1.7.

```
        Msg 3136, Level 16, State 1, Line 9
        This differential backup cannot be restored because the database
        has not been restored to the correct earlier state.
        Msg 3013, Level 16, State 1, Line 9
RESTORE DATABASE is terminating abnormally.
```

Figure 1.7 The error message

This is why you don't do full backups in the middle of the schedule: A full backup resets all the flags and throws off the schedule of the other types of backups.

However, SQL Server provides the copy backup as a way to do a backup that includes the entire set, just like a full backup, but does not reset the flags when it has completed. This is a good option if you need a backup but don't want to affect the existing backup schedule.

A great example for using a copy backup is when you need to do an unscheduled database refresh by taking a backup in production to use the database in a non-production environment for user-acceptance testing.

The copy backup will not upset the flag system since it ignores the flags when backing up and does not reset them once the copy backup is completed, as you see in Figure 1.8. You can make a safe copy without disrupting the week's backup schedule. You can run a copy backup in the middle of the week without resetting flags and disrupting the weekly differential or transactional log routines.

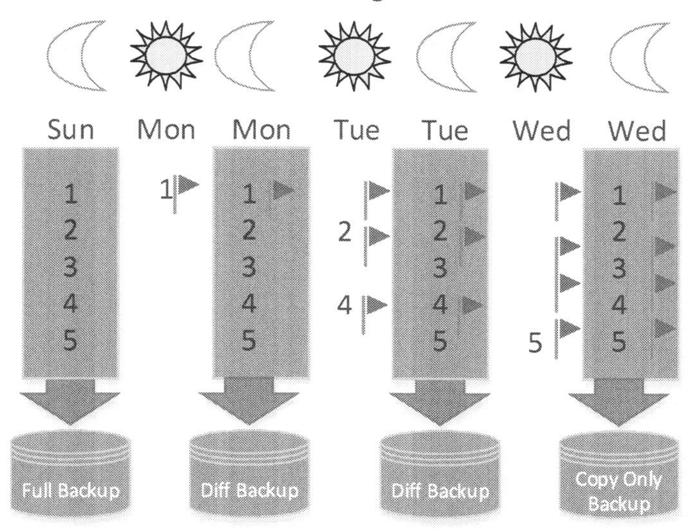

Figure 1.8 Flags not removed following a successful copy backup

Characteristics of Copy Backup

A copy backup has the following characteristics: It's like a full backup but doesn't reset the file flags. It is a safe option for an ad hoc backup during the week where you don't want to affect the current backup schedule.

Transactional Log Backup

The beginning of this chapter covered the purpose of the SQL Server database files. Recall that the transactional log records all the transactions and the database modifications made by each transaction. When changes occur, they are logged sequentially into the transactional log file, not the database file. A sequential unique identifier is associated with each change. The transactional log also breaks down the changes into buckets called virtual log files (VLFs).

Transactional log backups are key to having a recovery strategy that lets you recover up to point of failure. To leverage transactional log backups, you must have a base as your starting point. A proper base is a full backup or a full backup with the latest differential backup. You must restore transactional log backups in sequential order. Therefore, if you have a missing transactional log backup or a corrupt transactional log backup file, you will not be able to continue past the unrestorable backup file on to the next transactional log backup.

Compare Figure 1.9 with the previous figures illustrating backups. You'll see that the difference between the full backup and transactional log backup is that the transactional log backup will back up the data changes since the last log sequence number (LSN) was backed up. You can use a full backup as a base to start transactional log backups, but keep in mind that the full backup will back up all database pages.

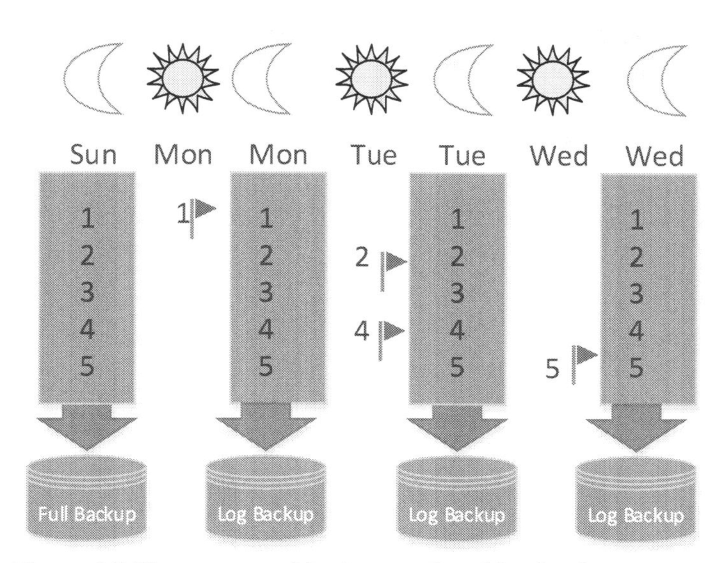

Figure 1.9 Flags removed by transactional log backups

Here's an example: On Sunday evening before the work week started, a full backup was performed. During the day on Monday, a transaction updated the first row of data and caused extent's flag to be raised, signaling that it needs to be included in the next transactional log backup.

On Monday evening, the transactional log backup includes the changes to the first row of data. The flag is now cleared, so the next transactional log backup does not include this row of data.

During the day on Tuesday, transactions updated rows 2 and 4. The flags were raised, so rows 2 and 4 are included in the next transactional log backup. On Tuesday evening the next transactional log backup occurs. This backup clears the flags for rows 2 and 4, which were updated during the day.

If a crash happened on Wednesday morning, you would need the full backup from Sunday and the log backups from Monday to recover the update to row one. Tuesday night's transactional log backup would also need to be applied to recover the updates to row 2 and row 4. Transactional log backups back up only the changes that have occurred since the last transactional log backup.

Transactional log backups require less storage space than differential and full backups, but the tradeoff is that transactional log backups are more intensive to restore because all transactional log backups are required to recover to the point of failure, as you can see in Figure 1.10.

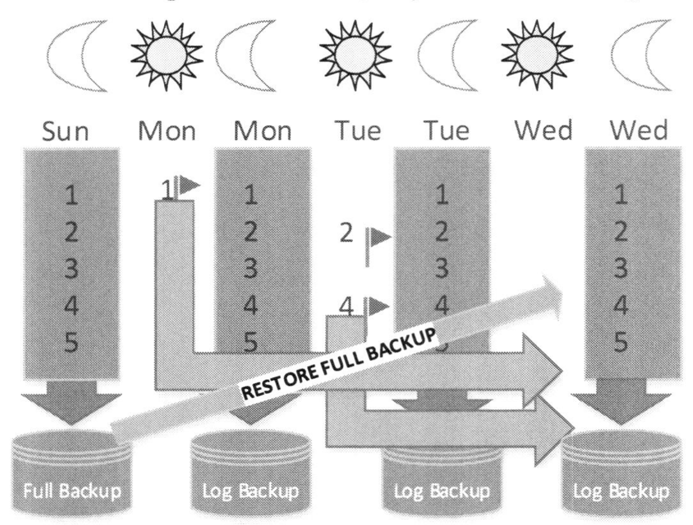

Figure 1.10 Restore process from full backup and multiple log backups

If you're still trying to wrap your head around how transactional log and transactional log backups work, the next few pages provide a step-by-step tutorial that follows data through the data file and logfile as it enters a new database. This example will cover how inserts, updates, deletes, and transactional log backups affect the transactional log.

Step by Step

Step 1. A brand new database has only one table (Leads), which contains zero records. No records are in the LinchpinPress database, so no records are in the data file. Since you have not made any changes to the database, zero records are in the logfile, too. If you ran the following query:

```
SELECT * FROM Leads
```

No records would be returned, as you see in Figure 1.11.

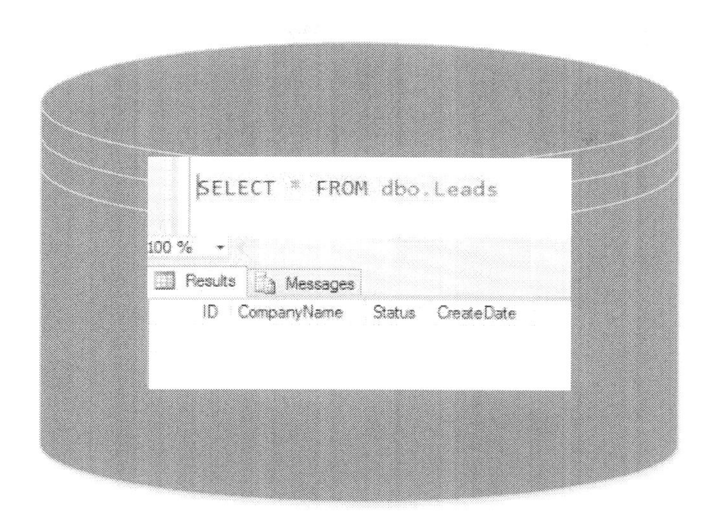

Figure 1.11 No records are in data file and no records are in transactional log file

Let's take a full backup to set up the base starting point for transactional log backups.

Step 2: Data starts coming into the LinchpinPress database. As Figure 1.12 shows, one new record for Linchpin People has been added in the Leads table. Now one record is in the data file, and one record is in the transactional log file.

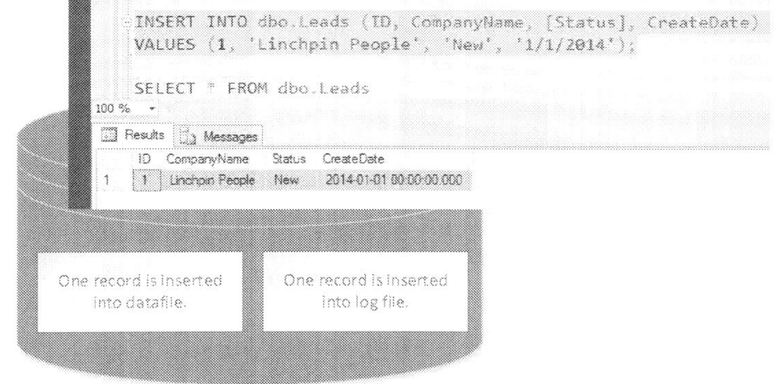

Figure 1.12 One record in data file; one record in the transactional log file

Step 3: Add another record into the leads table (Procure SQL), as in Figure 1.13. Two records are now in the LinchpinPress database, so two records are in the data file and two records are in the log file. Two entries in the log file reflect the inserted records.

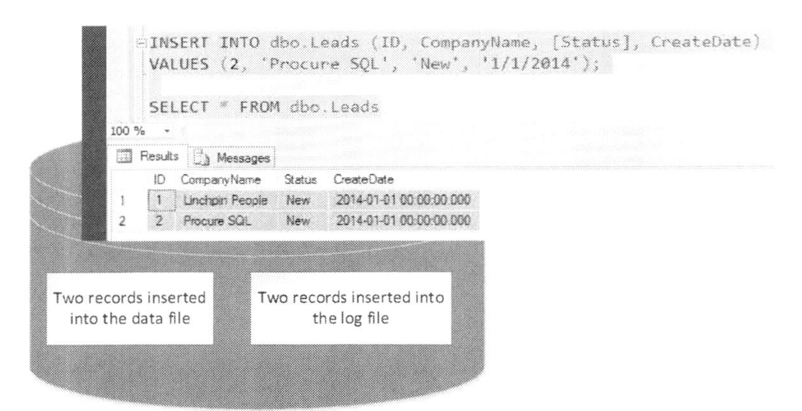

Figure 1.13 Two records in the Leads table; two records in the transactional log

Step 4. The next transaction updates an existing record. Lead 2 (Procure SQL) is a successful lead, and its status changes from New to Customer. Two records are still in the database, so that data file contains two records. Three records are now in the log file because you've made three changes (the two inserts and an update, as you see in Figure 1.14) since the last time you backed up the database.

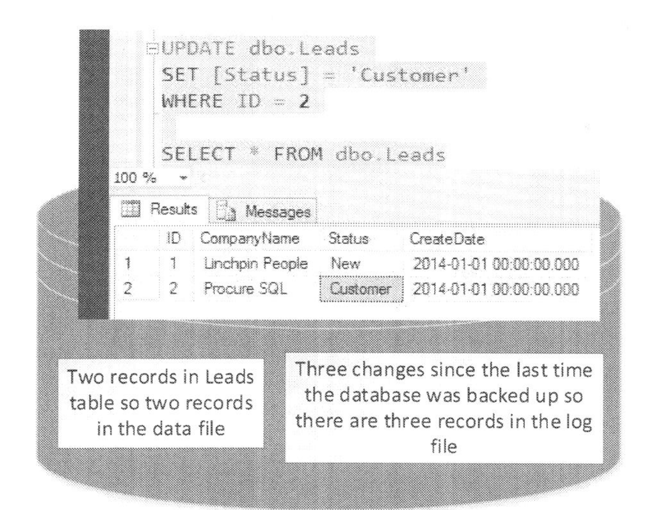

Figure 1.14 2 records in the Leads table, but 3 records in the transactional log

Step 5. The LinchpinPress database is backed up nightly at midnight. After the three changes happen, suppose it's after midnight and the transactional log backup has just finished running. At 12:15 AM, two records are still in the Leads table, and two records are in the data file.

During most backup processes, the changes contained in the logfile are sent to the backup file. The logfile is truncated as part of the backup process, so zero records remain in the logfile immediately after each transactional log backup, as Figure 1.15 shows. (How SQL Server handles this process will be discussed later.) Transactional log files only issue checkpoints to flush the records upon *logfile* backups. Regular backups do not impact the transaction log.

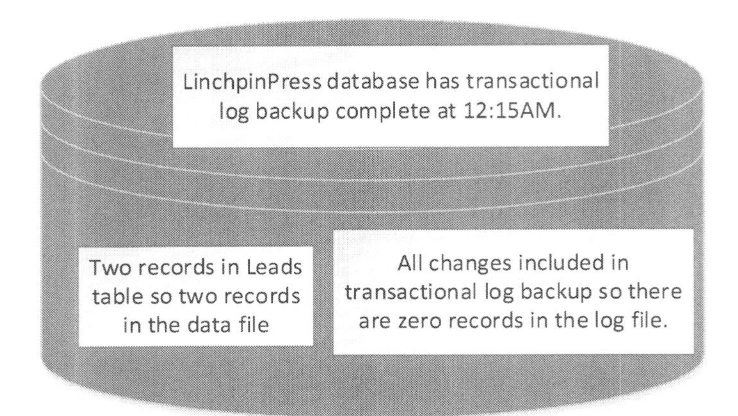

Figure 1.15 The log file is truncated, so it contains no records

Step 6. On the next day, you insert a Straight Path IT Solutions record (the third record added to the Leads table). Three records are in the data file. The transactional log file has been emptied since the backup. Therefore, this change now adds one change into the transactional log file, as you see in Figure 1.16.

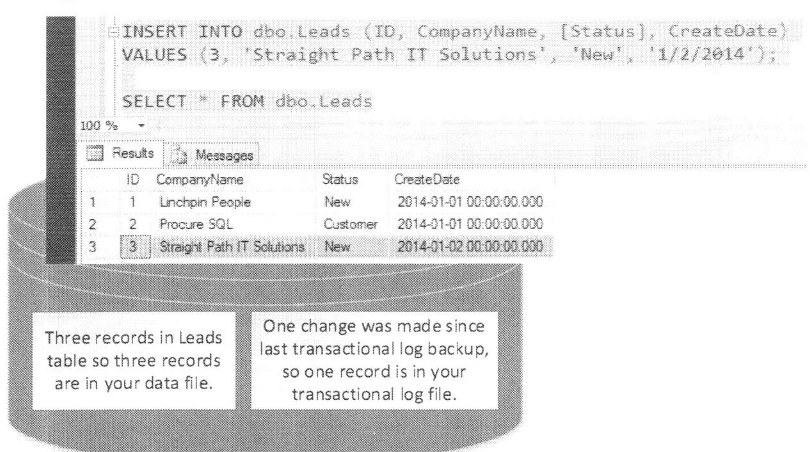

Figure 1.16 Three records in data file; one in the transactional log file

Step 7. On the same day (Day 2), you delete Procure SQL from the Leads table. Removing one record leaves two records in the Leads table and the data file. The transactional log file now contains two records because two changes have occurred since the last transactional log backup. These two changes are the INSERT (Straight Path IT Solutions) and the DELETE (Procure SQL), as shown in Figure 1.17.

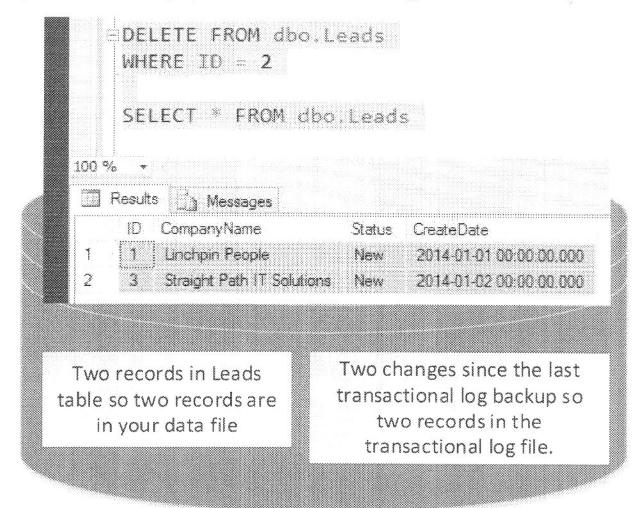

Figure 1.17 Record ID 2 (Procure SQL) deleted on day 2: two records in data file; two records in the transactional log file

Summary

The database is a direct reflection of all the data stored in the data files that make up the database. Data files often have the MDF or NDF extension. SQL Server allows you to choose where these data files are located. Data files change over time, so the database is also changing. The data files cannot tell what changed a minute ago versus what changed a year ago.

The transactional log files keep track of the latest changes to the database. Log files often have the LDF extension. Transactional log files continue to grow until they are truncated. If you back up your database with transactional log backups, then the log files are truncated at that time.

Figures 1.18 and 1.19 summarize and compare the backup types discussed in this chapter.

Backup Type	Backs up	Resets Marker
Full/Normal	Entire Set	Yes
Log Backup	Changes since last log backup	Yes
Differential	Only Marked Files	No
Copy	Entire Set	No

Figure 1.18 Side by side comparison of each backup type

Backup Type	Space Consumption	Recovery
Full/Normal	Most	Use 1 File
Log Backup	Least	Uses Many Files
Differential	Medium	Uses 2 Files
Copy	Most	Use 1 File

Figure 11.9 Side by side comparison of backup type system

The most robust backups are the full and copy because they include the whole database. However, they also use the most storage space because the whole database is backed up when you use these types of backup.

The copy backup is especially useful in a system test because it allows a full restore without interfering with scheduled differential backups. The copy backup can be maintained with as little as one backup file.

The most complete backup is the full backup. It backs up all files and resets the markers that signal changes to all altered files. The full backup is used for weekly backup and leaves a clean slate for the upcoming week. The full backup can be maintained with only one backup file.

The transactional log backup is a good choice for frequent backups. The benefit of the transactional log backup is that it uses less storage space than full backups or differential backups. The drawback, however, is that a transactional log backup can be cumbersome during the restore process: Each subsequent transactional log backup from the full backup has to be restored individually.

The differential backup is also a good choice for daily backups because it backs up only extents that have changed since the latest full backup. It uses more storage space than the transactional log backup but is much less cumbersome to restore. The differential backup can be maintained with as little as two backup files.

The differential backup can also be used to reduce the required amount of transactional log backups to restore to a point of failure. When you restore a full backup with latest differential backup, you only have to apply the transactional log backup files taken after the latest differential. This will allow you to skip multiple transactional log backups throughout the week.

Points to Ponder
File Backups

1. Backing up critical data is very important but useless unless you can restore the backups in the proper order.
2. A full backup backs up the entire set and resets the markers that were not reset by other backups.
3. The differential backup will back up only the files that have markers raised and will not reset the markers.
4. The copy backup is like a full backup that does not reset any of the markers that signal changes to an extent.
5. SQL Server stores data in one or more data files.
6. SQL Server stores changes in one or more transactional log files

Review Quiz – Chapter One

1. The policy is to perform a full backup every Sunday night and transactional log backups every weeknight. On Wednesday morning, the system is corrupted and needs to be restored. Which backups are needed?
    a. Just the latest full backup from Sunday
    b. Just the latest transactional log backup from Tuesday
    c. The latest full and the latest transactional log backup from Tuesday
    d. The latest full and both Monday's and Tuesday's transactional log backups

2. The policy is to perform a full backup every Sunday night and differential backups every weeknight. On Wednesday morning that the system is corrupted and needs to be restored. Which backups are needed?
    a. Just the latest full backup from Sunday
    b. Just the latest differential backup from Tuesday
    c. The latest full and the latest differential from Tuesday
    d. The latest full and both Monday's and Tuesday's differential backups
3. What types of backups do not reset the backup markers?
    a. Full
    b. Copy
    c. Transactional Log
    d. Differential
4. Which files get automatically truncated after a backup?
    a. Data files
    b. Transactional log files
5. What type of action will cause the transactional log file to grow but not the data file?
    a. SELECT
    b. INSERT
    c. UPDATE
    d. TRUNCATE
6. What type of action will cause the transactional log file and the data file to increase in size?
    a. SELECT
    b. INSERT
    c. UPDATE
    d. TRUNCATE

Answer Key

1. You want to restore up to Tuesday, so (a) only gets you to Sunday and is incorrect. A transactional log restore can only be done if the first restore is from a full backup, making (b) incorrect. The transactional log backups need to be done in order until the last one. Since there are

two transactional log backups, you need one full restore and two transactional log restores, making (d) the correct answer.

2. You want to restore up to Tuesday so (a) only gets you to Sunday and is incorrect. A differential restore can be done only if the first restore is from a full backup, making (b) incorrect. For differential restores, you need to use only one of them (the most recent), making (c) the correct answer.
3. The full backup resets the markers, making (a) incorrect. The differential and copy backup do not reset the markers, making (b) and (d) both correct.
4. Data files hold all the database data even after a backup, so (a) is incorrect. After a transactional log backup, the transactional log file moves its data to the transactional log backup file and is truncated, making (b) the correct answer.
5. The SELECT statement does not change the data, so neither the data file nor the log file grows, making (a) incorrect. INSERT will cause both the data and log files to grow, making (b) incorrect. TRUNCATE does not affect the transactional log file. If you update a record, the record could be the same size but each update is logged, making (c) correct.
6. If you insert a record that adds data to the database, this action is also logged making, (b) the correct answer.

Chapter 2. Performing a Full SQL Server Backup

Full backups are the key to any backup strategy. Since a full backup is all-encompassing (as Chapter 1 explained), it includes everything needed to recover the database to the state it was in at the time the full backup was taken. That's why a full backup is also the foundation of any restore sequence.

A full backup is what makes the smaller and faster incremental backups and differential backups possible. This is because the very first incremental backup or differential backup looks to see what has changed since the last full backup. Therefore if you've never done a full backup, you can't do a differential or incremental backup.

How to Do a Full Backup

Taking a full backup is one of the easiest tasks you can perform as a data professional. Have you ever copied a file to a disk or USB drive? Most of the time, you use a drag-and-drop operation in Windows Explorer to do this. You may also be aware that you can use the command line utility (which looks like DOS) to copy that same file to your disk or USB drive. Either way you do this, you get the same results: The file is copied and saved in a different location where you can access the backup if you need it. To copy your SQL Server databases, you can easily use SQL Server Management Studio's (SSMS's) graphical UI or T-SQL code.

Many say that the most important role for a database professional is to ensure that adequate backups are on hand. Regularly scheduled full backups should be a part of your organization's maintenance for production databases. In case of a disaster, if you can't recover applications and web and database servers, you can probably rebuild them. However, it is nearly impossible to recreate data if you can't recover it.

> **NOTE:** For the exercises in this chapter, you need to create a folder called Backups on the local C: drive (C:\Backups). You will also need a recent copy of the sample AdventureWorks database, which you can download at *http://msftdbprodsamples.codeplex.com/releases/view/93587*. In this book we are using AdventureWorks 2012, which we renamed to AdventureWorks. You will need to have run the SetupScript01.sql file. You can find all scripts mentioned in this chapter in the Book Series section at *www.LinchpinPress.com*.

To make a full backup using the SSMS graphical UI, you need to follow a few simple steps. First, right click on the database and choose **Tasks** >. Select the **Back Up…** option (as Figure 2.1 shows). Do this now with AdventureWorks.

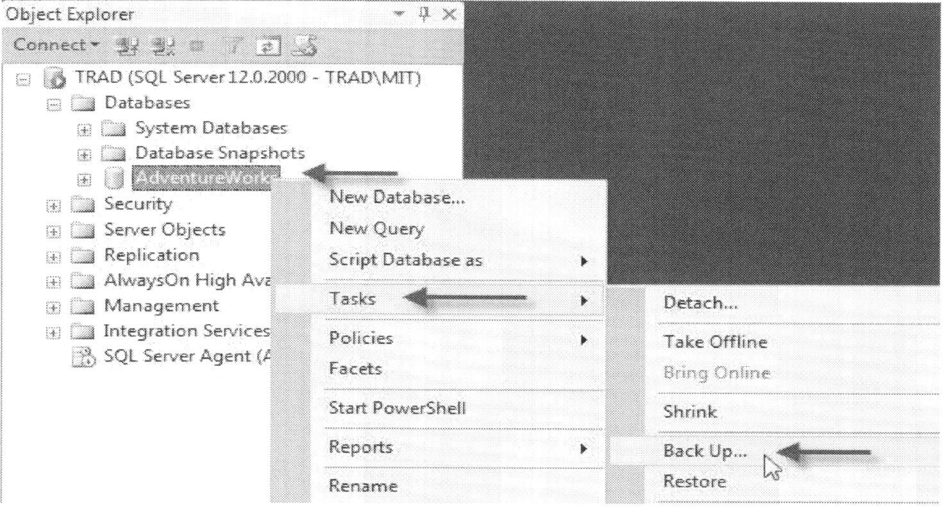

Figure 2.1 Right click on the database, choose Tasks, and then Back Up…

Once you have chosen **Back Up…** , the general dialog box appears. From this dialog, you can select from several options. Selecting **OK** at this point creates a full backup of the database into the default backup location for the SQL Server instance.

In Figure 2.2, the first option is to choose which database to back up.

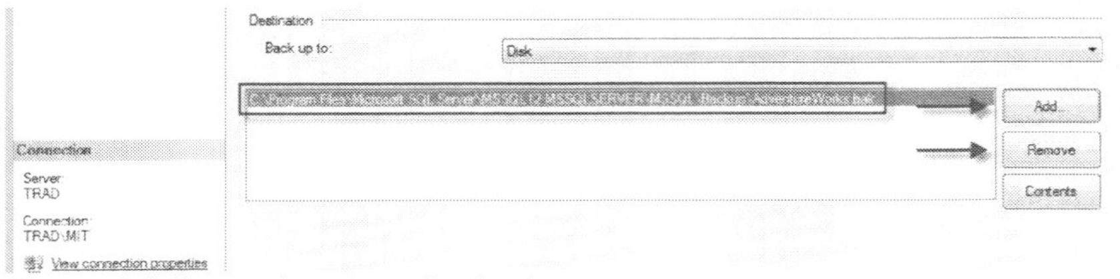

Figure 2.2 General options for backup

Next is the backup type you want to make. In this case, you want to choose **Full**. For **Backup component**, make sure **Database** is selected. The next item is the destination and name of the backup file.

The default path is set at the instance level. For SQL Server 2014, the default location is the following Path:

C:\Program Files\Microsoft SQL Server\MSSQL12.MSSQLSERVER\MSSQL\Backup\

In this example, you do not want to save in that location. To change this for the example, select the path to highlight it and then click **Remove** >. Then click **Add** so you can specify a new path.

To pick your own location, enter the path and name of the database backup file to be created (as shown in Figure 2.3).

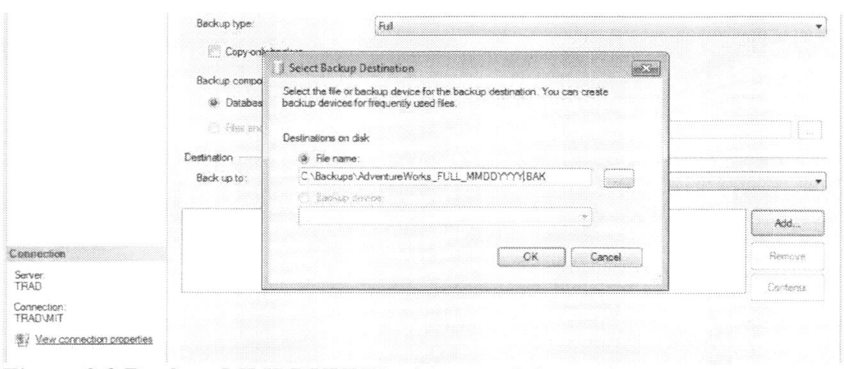

Figure 2.3 Replace MMDDYYYY with actual date value

A best practice is to include a date and time, as well as backup type, in the backup name. In this example, change the backup file name to **AdventureWorks_MMDDYYYY.BAK**. Type the path and name of the database file. Replace MMDDYYYY with actual date value in the form of Month Day Year. For example, if today is June 1, 2014, then the filename should be AdventureWorks_Full_06012014.BAK. This naming convention gives you a nice visual aid so you don't have to look at the timestamp on the file itself. Click **OK**.

NOTE: When creating a custom backup job, you would normally want to include HHMMSS in the name, as well.

The choices so far in the **General** page of the **Back Up Database** dialog are just a few of the options available. Just below **General**, click **Media Options** to see several more possible backup choices. Figure 2.4 shows that the default values are already checked.

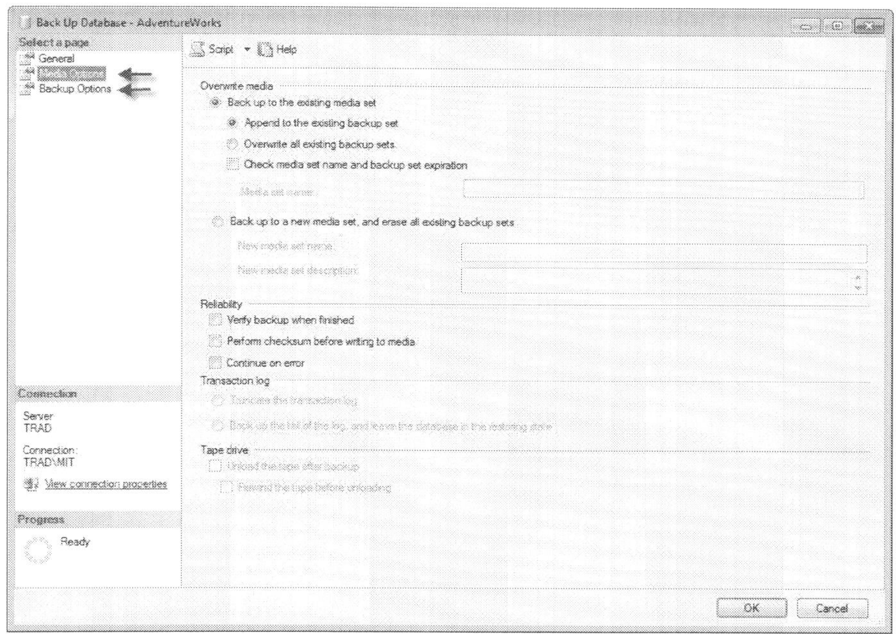

Figure 2.4 Additional options for backups

Odds are pretty good that someone will perform many backups of the same database over time. If you back up the database 10 times, does that mean you have 10 backup files? Or can you put 10 backups into 1 single file? The answer is you get to choose.

If you chose the option to append backups and use the same filename each time, then that one filename will hold multiple backup sets. Be careful not to use the same name each time unless you are using the append option, otherwise you will overwrite the old backup and have only the latest backup saved.

Overwrite media defaults to **Append to the existing backup set**. This was chosen to maintain backup history by putting many backup sets into one backup file. If you make multiple backups and use the same file name each time, multiple backup sets would be within that single backup file. If you want to overwrite the old backup with the latest backup of the same name, you have to select the **Overwrite all existing backup sets** option.

Make Sure It Works

Why is it a good idea to have a spare key to your car? I bought a new car six months ago and had a spare key made. On a very important day, I found out I should have tested the spare. The key did not work, and my original key was lost. After a very expensive house call from the locksmith to recreate my key's data, I learned to test my backup key and not just assume it will work.

This story leads to the topic of reliability, which is the next section of the full backup. A false sense of security may come with choosing some of the reliability options.

The recommended reliability option is **Verify backup when finished.** However, performing this verification does not mean the backup file is 100 percent valid. The only true way to fully verify a backup is to restore it. Everyone should have a process in place to regularly test backups by performing restores.

Compression

If you have ever packed for a long trip, you've discovered that you can pack more items into a suitcase by compressing your clothes and reducing the amount of space they take up. By reducing the amount of wasted space in your suitcase, you are more efficient at carrying more items, therefore reducing the need to carry an additional bag.

SQL Server 2008 Enterprise and newer can also utilize compression. Backing up using compression means that backups will take up less space and also take less time to complete because they require less device I/O.

If you are using a version of SQL Server that supports compression, you will see that option listed. You can use one of the three options:

- Use the default server setting
- Compress backup
- Do not compress backup

Let's use the default option, which is shown in Figure 2.5.

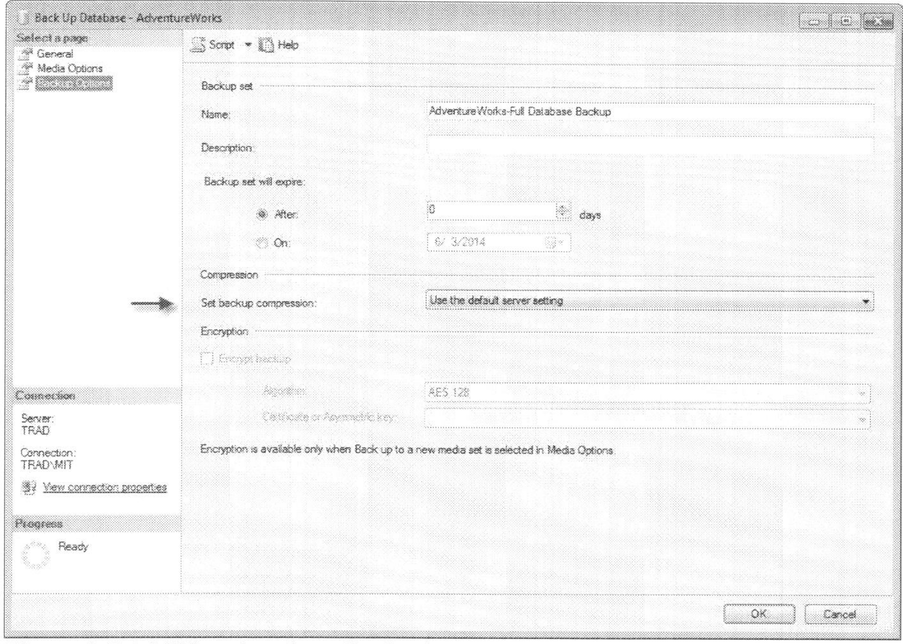

Figure 2.5 Additional options for backups

Click OK, and the AdventureWorks database will back up to:
C:\Backups\AdventureWorks_Full_MMDDYYYY.BAK.

Using T-SQL

You have many choices and options for making a backup. If use the UI and you click all the right buttons and do everything perfectly, then you will likely be asked to do it again many times. Forgetting one checkbox just one time, can cause the wrong type of backup, which could be unusable. Or worse yet, you could accidently overwrite a file you may need later.

If you could lay out this perfect set of steps into a T-SQL script, then you'd just need to run that script at the right time. This way significantly reduces the chance for human error. Therefore the preferred choice for making backups is to use T-SQL code.

The syntax to back up with T-SQL code is very straightforward:

```
BACKUP DATABASE DB_NAME TO <backup_device> WITH <options>
```

In this example, let's name the backup so it's easy to tell that T-SQL was used to create this file. To do this, append _TSQL to the file name. This is different from the file name created using the graphical UI in the previous section.

Type the path and name of the database file. Replace MMDDYYYY with actual date value in the form of Month Day Year. For example, if today is June 1, 2014, then the filename will be AdventureWorks_Full_06012014.BAK.

Let's go ahead and make a full backup by using the following script:

```
BACKUP DATABASE AdventureWorks TO DISK =
'C:\Backups\AdventureWorks_FULL_MMDDYYYY_TSQL.BAK'
```

The message in Figure 2.6 shows that the backup was successful.

```
Messages
Processed 24240 pages for database 'AdventureWorks', file 'AdventureWorks2012_Data' on file 1.
Processed 3 pages for database 'AdventureWorks', file 'AdventureWorks2012_Log' on file 1.
BACKUP DATABASE successfully processed 24243 pages in 8.278 seconds (22.879 MB/sec).
```

Figure 2.6 Shows full backup was successful

When using the graphical UI, you have lots of options to choose from, such as append, overwrite, verify backup, and many more. When you use T-SQL, many of these options are also available when you specify the WITH clause.

One handy feature of a graphical UI backup is that a display box shows the percentage of the database that has been backed up. To show the percentage backed up when you're running T-SQL code, you will need to specify

```
WITH STATS = X
```

(with X being the increment of percentage). So if you specify WITH STATS = 10, the backup process will display percentages from 0, 10, 20 ...100 as it completes. Most DBAs use a 1 or 10 as percentage increments.

To use compression, select WITH COMPRESSION in the syntax. If you want to overwrite the existing backup file, specify WITH INIT in the syntax.

Using the Graphical UI to Restore

Using the graphical UI to restore is much like using the graphical UI to back up the database. As you see in Figure 2.7, you start the restore by right clicking the database (e.g., AdventureWorks) and choosing **Tasks** > **Restore** > **Database**.

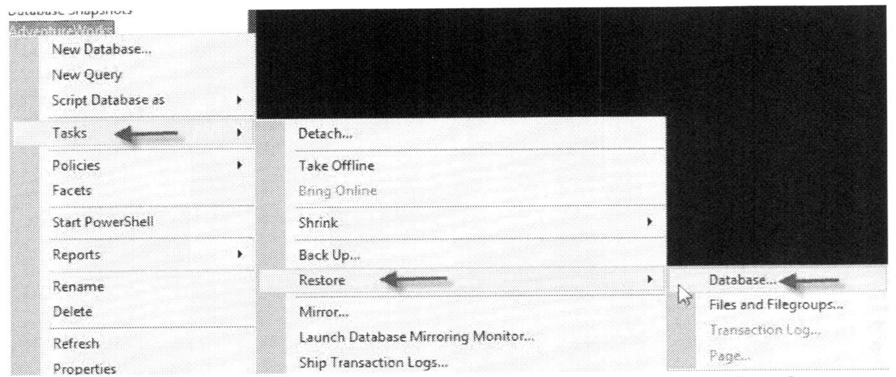

Figure 2.7 Right click on the database, chose Tasks > Restore > Database

A new window will open on the **General** page, prompting the choice of the database source and destination.

Since you typically restore database files from one server to another (such as production to a test server), you will choose **Device** as the **Source** and then click the ellipsis (as shown in Figure 2.8).

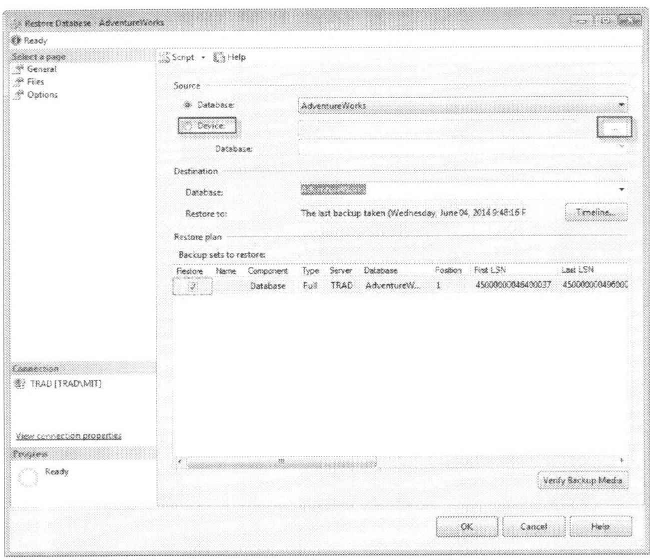

Figure 2.8 Click Device and then the ellipsis

Where is that backup file located? SQL Server will want to know this, and here is the chance to tell it exactly where that file was saved. You need to select your backup file to restore your database and add that exact path to SQL Server. In Figure 2.9 the Backup media type is file. Then click Add.

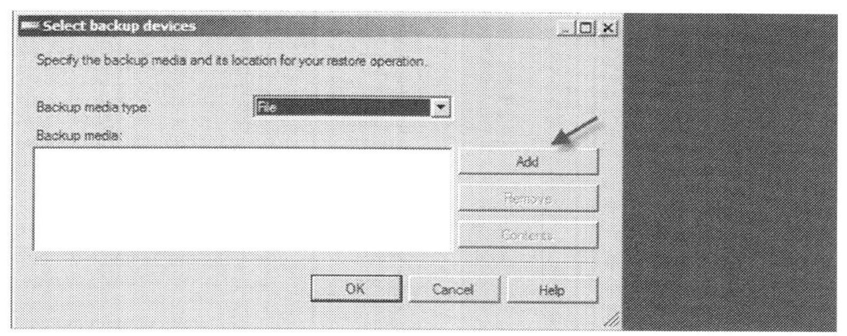

Figure 2.9 Click Add

The file was saved to the C drive in the backup folder. So for this demo, browse to the C: \Backups folder to see both files that were created. You can select either backup file since they are the same. Select the first one and chose **OK** (as in Figure 2.10) and chose **OK** again.

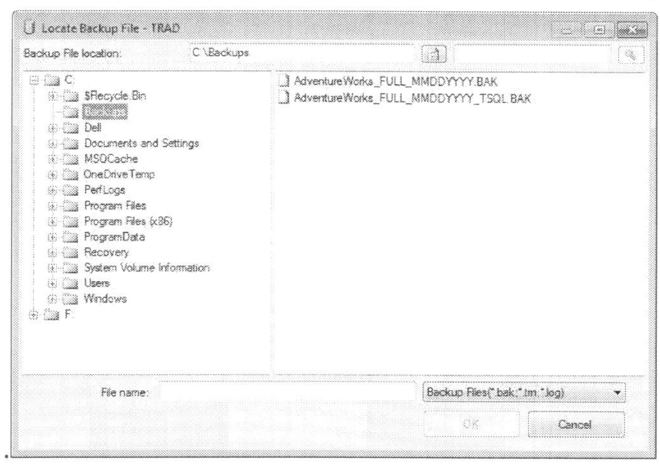

Figure 2.10 Browse to C:\Backups, select the file and click OK and OK again

This returns you to the general page of the **Restore Database** window. Click on the **Files** page to change the location of where to restore the data and log file. You may have to restore a database to a server that does not have exactly the same disk configuration. Or you may want to restore the files to a location other than the default for that instance (see Figure 2.11).

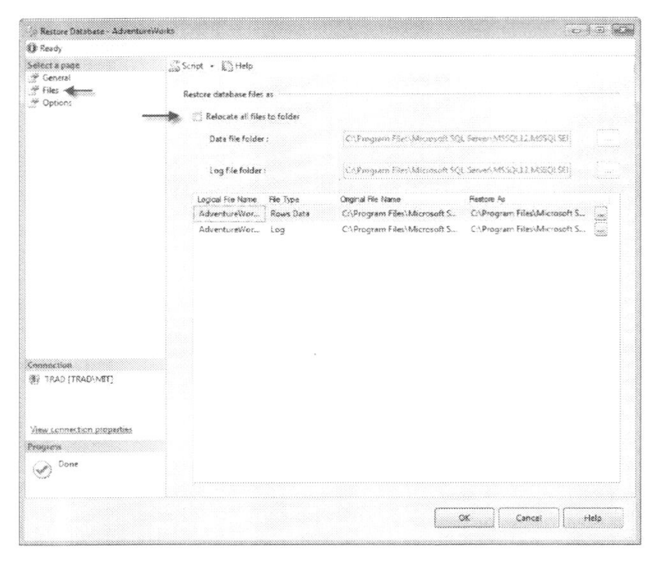

Figure 2.11 Change the location to restore the database files

Now click on the **Options** page (as in Figure 2.12). You have several restore options to choose from. Check the box to **Overwrite the existing database (WITH REPLACE)**.

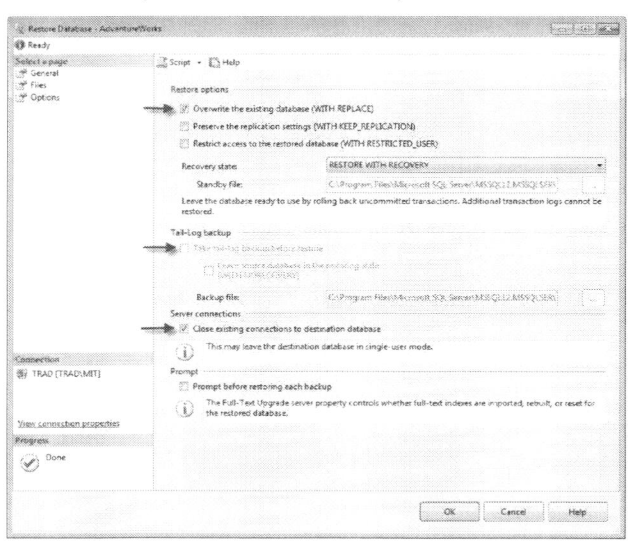

Figure 2.12 Check to Overwrite, uncheck the box to back up the tail-log. Check to Close existing connections to destination database

Some restores will come from the full backup. But you may need to restore a differential afterwards. So know ahead of time whether this is all of your data or whether you have a differential backup to do to add in more data. In this example, you only took this full backup and no other type of backup. Since this is the only restore before making the database live, set the **recovery state** to **RESTORE WITH RECOVERY**. By choosing to restore with recovery, you are setting the database to be open and running as soon as this restore is completed.

As you will see in future chapters, sometimes you will need to leave the database offline so that additional restores can be applied. Next make sure the box to **Take tail-log backup before restore** is unchecked. This particular option in the restore dialog was new to SQL Server 2012. Chapter 4 will cover tail-log backup terminology.

Nobody should be connected to or using this database until you are done with the restore. Some system connections might have to be told to disconnect during this restore. You can close any existing connections to the database by checking the box **Close existing connections to destination database**. I want to point out that I typically like to check for existing connections by using the SP_WHO2 system stored procedure to ensure I am not affecting any active users.

You are now ready to restore the database, so click **OK**. Congratulations! With this process you have successfully restored the AdventureWorks database (as Figure 2.13 confirms).

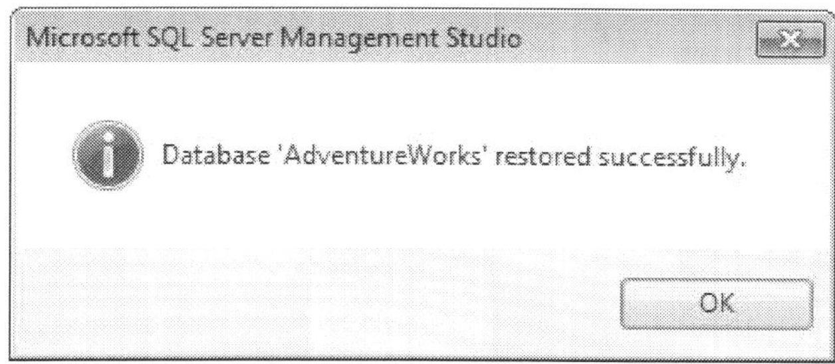

Figure 2.13 Confirmation window that the database was restored

Using T-SQL to Restore

The benefit to using T-SQL code is once you get it right, you can run the code any time. If you use the point-and-click method, you need to make your choices each time you run the backup. T-SQL scripts also let you save a generic script that has all the syntax you usually use so you don't have to retype it each time you need to restore a database. All you have to do is make a few changes for that restore and execute the script.

The syntax for a database restore using T-SQL is

```
RESTORE DATABASE DB_NAME FROM <backup_device> WITH
<options>
```

You have several options you can use after the WITH statement. You can specify the recovery setting RECOVERY / NORECOVERY / STANDBY

In this case, you will be fully recovering the database so users can connect by specifying WITH RECOVERY.

You can also use WITH MOVE to specify the location of the database files, or REPLACE to overwrite the existing database. As stated earlier, you can see the progression of the restore by using STATS. Type the path and name of the database file. Replace MMDDYYYY with the actual date value in the form of Month Day Year. For example, if the backup was created on June 1, 2014, then the filename could be C:\Backups\AdventureWorks_Full_06012014.BAK:

```
USE [MASTER]
RESTORE DATABASE [AdventureWorks]
FROM DISK = 'C:\Backups\AdventureWorks_Full_MMDDYYYY.BAK'
WITH MOVE 'AdventureWorks2012_data'
TO 'C:\Program Files\Microsoft SQL
Server\MSSQL12.MSSQLSERVER\MSSQL\DATA\
AdventureWorks2012_data.mdf',
MOVE 'AdventureWorks2012_log'
TO 'C:\Program Files\Microsoft SQL
Server\MSSQL12.MSSQLSERVER\MSSQL\DATA\
AdventureWorks2012_log.ldf',
STATS = 1, REPLACE
```

> **NOTE:** The code above is a likely location for the AdventureWorks MDF and LDF file(s). You can always get the properties of AdventureWorks and select the **files** page to see where the database storage is located.

Summary

Full backups are the key to any backup strategy. Full backups include everything needed to recover the database. You can easily use the graphical UI or T-SQL to make and restore full backups. Having a regularly scheduled full backup is critical to protecting your organization's data assets. You can easily accomplish this by using either the built-in database maintenance plan or creating our own custom SQL Agent job. The only way to fully validate that your backups are good is to regularly restore them.

Points to Ponder—Full Backup

1. Full backups contain the entire database, including all files and file groups associated with the database.
2. A full database backup provides a complete copy of the database.
3. Backups do not cause blocking, contrary to any myths out there. However, backups are very I/O intensive, which can cause performance issues related to I/O if you run backup during high peak times.

Review Quiz – Chapter Two

1. A full backup consists of which items?
    a. All transactions since the last full backup.
    b. Only changed data since the last full backup when the flag is not reset.
    c. Everything needed to fully recover the database.

   d. Only the primary file group.
2. When restoring a database, which recovery setting should you use to recover the database while leaving it in a usable state?
   a. NORECOVERY
   b. STANDBY
   c. ONLINE
   d. RECOVERY
3. When restoring a database using T-SQL, which command would you use to over write the existing database?
   a. WITH OVERWRITE
   b. WITH REPLACE
   c. WITH FORCE
   d. WITH COMMIT

Answer Key

1. A full backup contains all data needed to fully recover the database, including all file groups and enough of the transaction log to fully recover the database. Therefore answer (c) is correct.
2. To restore the database and set it so that online connections can connect, you need to recover the database. Answer (d) is correct.
3. When you choose to use T-SQL to overwrite an existing database during a restore, you must use the WITH REPLACE command. Using the graphical UI, you select the box to overwrite the existing database. Answer (b) is correct.

Chapter 3. Differential SQL Server Backups

When you clean your kitchen after dinner and wash the dishes, you don't empty your cabinets and wash every dish. You just wash the dishes that you've dirtied. Similarly, when you back up data, if you only want to back up the data that has changed, you can make a differential backup, as Chapter 1 explained. Differential backups back up all the changed data since the last full backup. A differential backup does not reset the differential bitmaps (i.e., flags) that indicate whether the data has been backed up.

Implementing a backup strategy that utilizes both full and differential backups can drastically reduce your backup time and the space required to store historical backups. In some cases, switching to a weekly full and daily differential backup can make a difference of backing up 2.5 TB weekly to less than 2 GB.

Consider how much of your database is made up of static data that doesn't change versus how much data is being updated or added. If you are performing daily full backups, you are backing up the same unchanged data every single day.

The advantage to taking full backups every night is that the restore strategy is simple. Just restore the latest backup.

Implementing a backup strategy that takes advantage of both full and differential backups alters that restore strategy. It adds a step to the restore process. You must first restore the full backup (base) with NORECOVERY so you can then restore the differential backup with RECOVERY.

Since most differential backups are small, the additional time to restore them usually amounts to minutes. This slight increase in restore time is usually an acceptable tradeoff for the decrease in the amount of time and disk space required for backups. Experience proves that the additional step to restore a differential backup only adds minutes to the restore time, which still fits well within service level agreements.

NOTE: For the exercises in this chapter, you need to create a folder called Backups on the local C: drive (C:\Backups). You will also need a recent copy of the sample AdventureWorks database, which you can download at http://msftdbprodsamples.codeplex.com/releases/view/93587. This book uses AdventureWorks 2012, which is renamed to AdventureWorks. You will need to have run the SetupScript01.sql file. You can find all scripts mentioned in this chapter in the Book Series section at www.LinchpinPress.com.

Since differential backups can only be made after a full backup, reuse the full backup script from Chapter 2. Type the path and name of the database file. Replace MMDDYYYY with the actual date value in the form of Month Day Year. For example, if the backup was made on June 1, 2014, then the filename could be 'C:\Backups\AdventureWorks_Full_06012014.BAK':

```
BACKUP DATABASE AdventureWorks TO DISK =
'C:\Backups\AdventureWorks_FULL_MMDDYYYY_TSQL.BAK'
```

Using the Graphical UI

To make a differential backup using the graphical UI within SSMS, you need to follow a few simple steps—just like you did for the full backup: First, right-click on the database you want to back up, and choose **Tasks** > and the **Back Up**… option. Do this now with AdventureWorks (as Figure 3.1 shows).

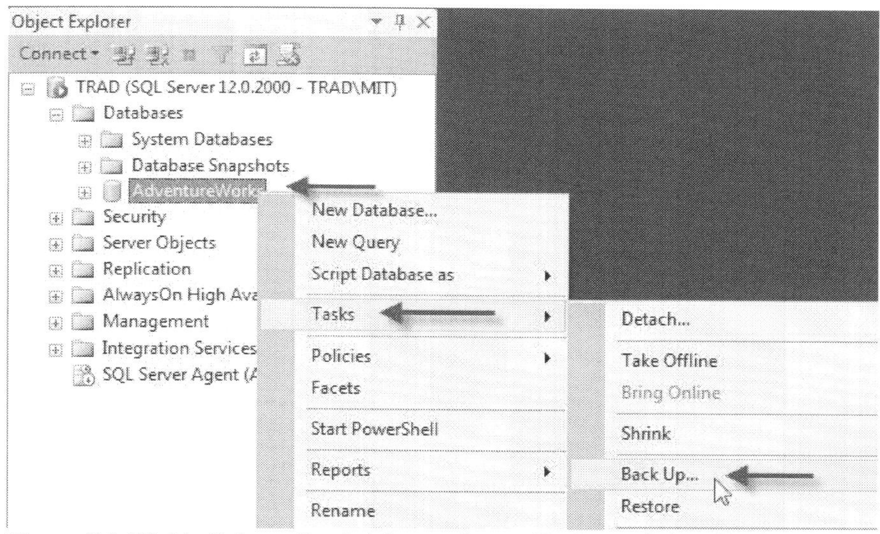

Figure 3.1 Right click on the database, choose Tasks, and then Back Up…

Once you have chosen **Back Up…** , the general dialog box appears. Here, you can choose from a number of selections. In Figure 3.2, the first option is to choose the database you want to back up.

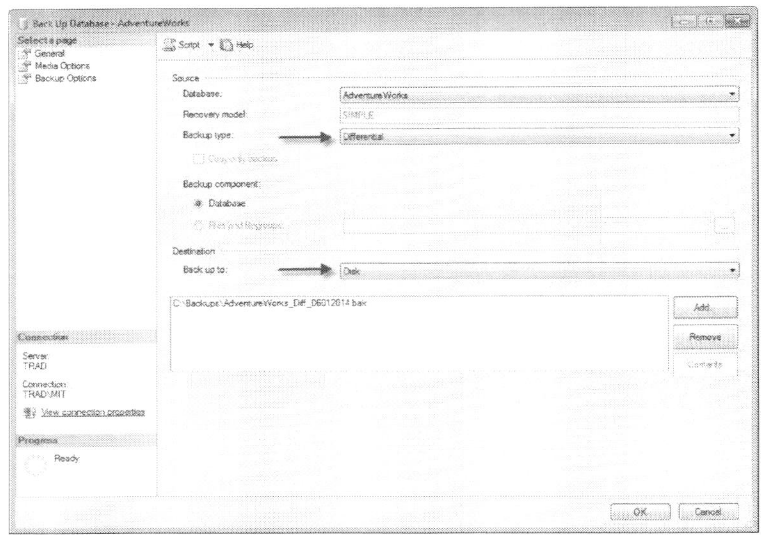

Figure 3.2 General options for backup

Next is the backup type you want to make. In this case, you want Differential. For **Backup component**, ensure that **Database** is selected. The next item is the **Destination**. Chose **Disk**. (URL backups will be covered in Chapter 10.)

The default path is set at the instance level. For SQL Server 2014, the default location is: C:\Program Files\Microsoft SQL Server\MSSQL12.MSSQLSERVER\MSSQL\Backup\. You want to change the path. So for this example, click **Remove** and then click **Add**.

Type the path and name of the database backup file (as you see in Figure 3.3). As with full backups, the best practice is to include a date and time, as well as the backup type in the backup name. Type the path and name of the database file. Type the path and name of the database file. Replace MMDDYYYY with the actual date value in the form of Month Day Year. For example, if today is June 1, 2014, then the filename should be AdventureWorks_DIFF_06012014.BAK. This gives you a nice visual aid so that you don't have to look at the timestamp on the file itself. Now click **OK**.

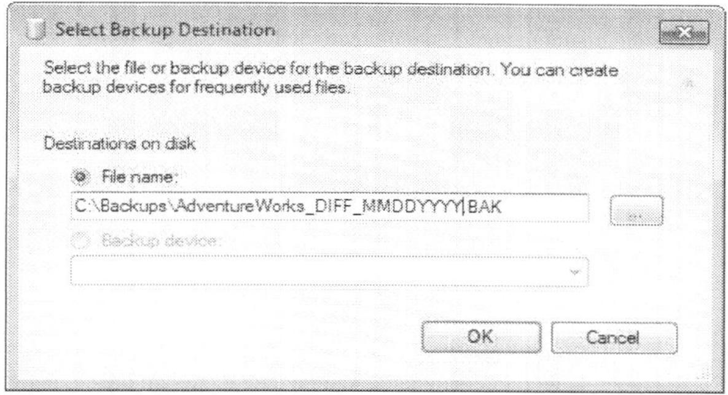

Figure 3.3 Replace MMDDYYYY with actual date value in the form of Month Day Year

When creating a custom backup job, if this is a scheduled job, include HHMMSS in the name, as well.

Click the **Media Options** page to see that you have several more options, as you did in Chapter 2 on full backups. Figure 3.4 shows the default values.

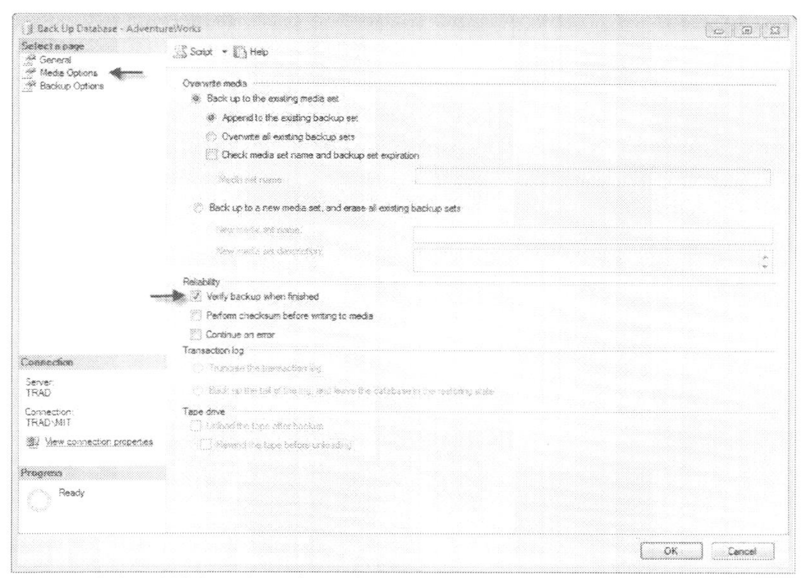

Figure 3.4 Additional options for backups

The **Overwrite media** option defaults to **Append to the existing backup set**.

Just like with full backups, you must verify differential backups by restoring them regularly. **Verify backup when finished** is still recommended, however that does not mean the backup file is 100 percent valid. If your backup solution includes full and differential backups, the restore validation should also include restoring from full and differential backups.

Click **Backup Options** page. Figure 3.5 shows the default values. Compression and encryption are also options with differential backups, and you perform them the same way as with full backups. Use the default option here, as you did in the previous chapter. Click **OK**, and the AdventureWorks database will back up to C:\Backups\AdventureWorks_DIFF_MMDDYYYY.BAK.

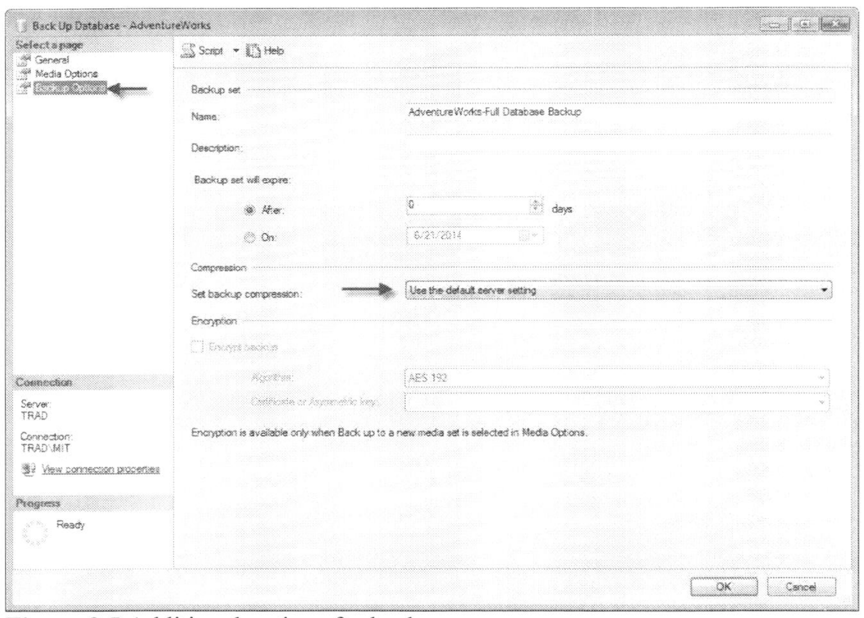

Figure 3.5 Additional options for backups

Using T-SQL

Chapter 2 covered the benefits of using T-SQL, explaining how having to check so many boxes each time you do a backup can be cumbersome. You can also say that anything worth repeating should be scripted. Backups are certainly worth repeating, so script them out for easier and more consistent reproduction. For that reason, the preferred choice for making backups is to use T-SQL.

The syntax is very straightforward. You specify:

```
BACKUP DATABASE DB_NAME TO <backup_device> WITH <options>
```

Then specify your options. Since you are making a differential backup, you have to specify WITH DIFFERENTIAL.

To give this file a different name from the file in the UI example, mark this backup file name to include _TSQL. This way you have the backup files used from the UI, along with the new ones you are creating now. Retain all backup files from this chapter.

Type the path and name of the database file. Replace MMDDYYYY with actual date value in the form of Month Day Year. For example, if the backup was made on June 1, 2014, then the filename could be 'C:\Backups\AdventureWorks_DIFF_06012014.BAK':

```
BACKUP DATABASE AdventureWorks TO DISK =
  'C:\Backups\AdventureWorks_DIFF_MMDDYYYY_TSQL.BAK'
WITH DIFFERENTIAL
```

Using the Graphical UI to Restore

Using the graphical UI to restore a differential backup is similar to restoring a full backup. As a matter of fact, your first step is to restore the full backup, as you did in Chapter 2.

To restore a differential database, you must first restore the prior full backup that the differential belongs to. You will need to follow the steps in Chapter 2. However, in the graphical UI, you specify the equivalent **RESTORE WITH NORECOVERY** on the options page.

Fortunately, when restoring multiple files (such as your full and differential backup), you can use the graphical UI to chain them together. To get started, right click on the database and choose **Tasks** > **Restore** > **Database** (as you see in Figure 3.6).

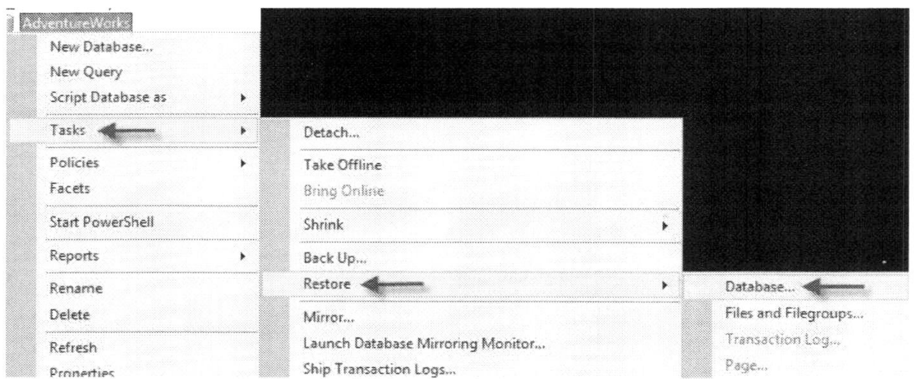

Figure 3.6 Right click on the database; chose Tasks > Restore > Database

A new window will open on the **General** page, prompting the choice of the database source and destination. You can restore from the backup files you have or restore from one live database to another without taking a backup. Since you typically restore database files from one server to another (such as production to a test server), you need choose **Device** as your **Source** and then click the ellipsis (as shown in Figure 3.7).

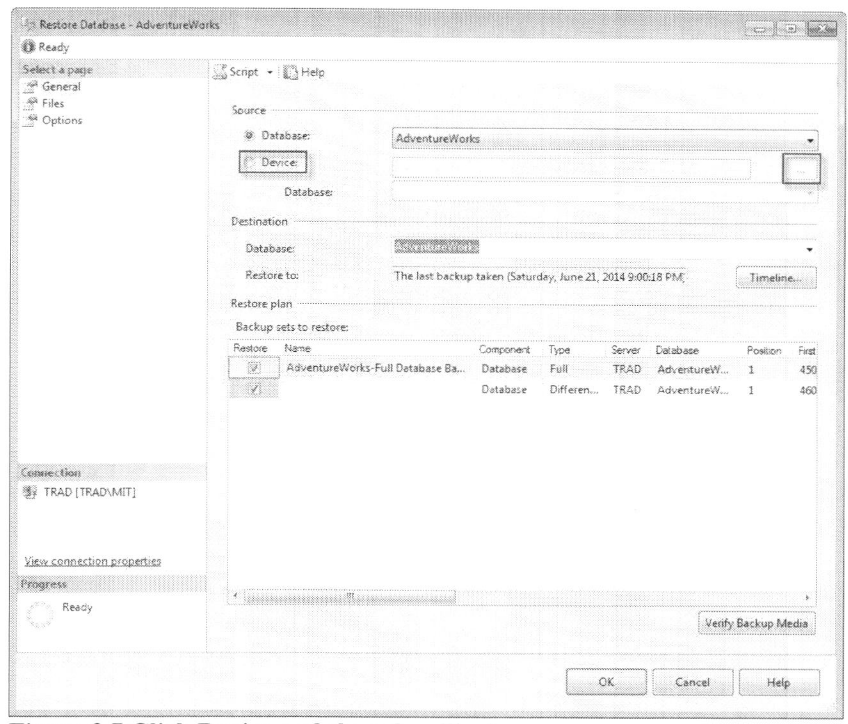

Figure 3.7 Click Device and then the ellipsis

Where are those backup files located? SQL Server will want to know this, and here is the chance to specify exactly where those files are saved. You need to select your backup file to restore your database and add that exact path to SQL Server.

As you see in Figure 3.8, your Backup media type should be **Files**. Then click **Add**.

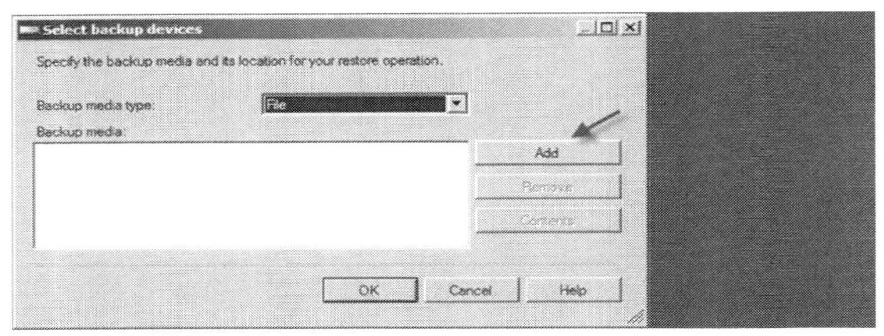

Figure 3.8 Click Add

Your file was saved to the C drive in the backup folder. For this demo, just browse to the C: \Backups folder to see your backup files that you created. You can then select your full backup file and the differential that belongs to the full. Chose the ones you used during the UI backup. Chose **OK** and then **OK** once more (as shown in Figure 3.9).

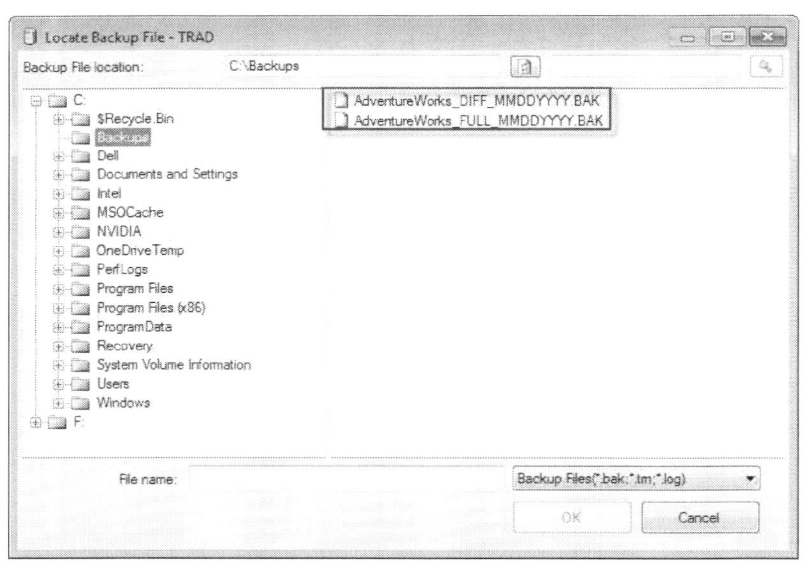

Figure 3.9 Browse to C:\Backups, select the files, and click OK

This returns you to the **Restore Database** window. Click on the **Files** page to see that you can change the where you want to restore the data and log file. In some situations, you may have to restore a database to a server that does not have exactly the same disk configuration. Or you may want to restore the files to a location other than the default location for that instance (see Figure 3.10).

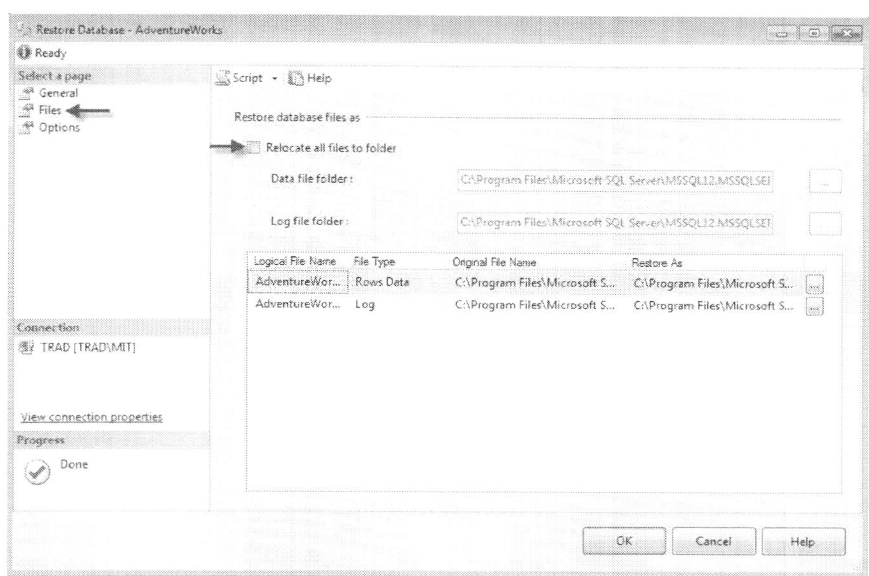

Figure 3.10 The Files page is where you can change the location to restore the database files

Now click on the **Options** page (which you see in Figure 3.11). On this page, you have several restore options to choose from. Check the box to **Overwrite the existing database (WITH REPLACE)**. This lets you leave the database offline so that additional restores can be applied, or you can bring the database online.

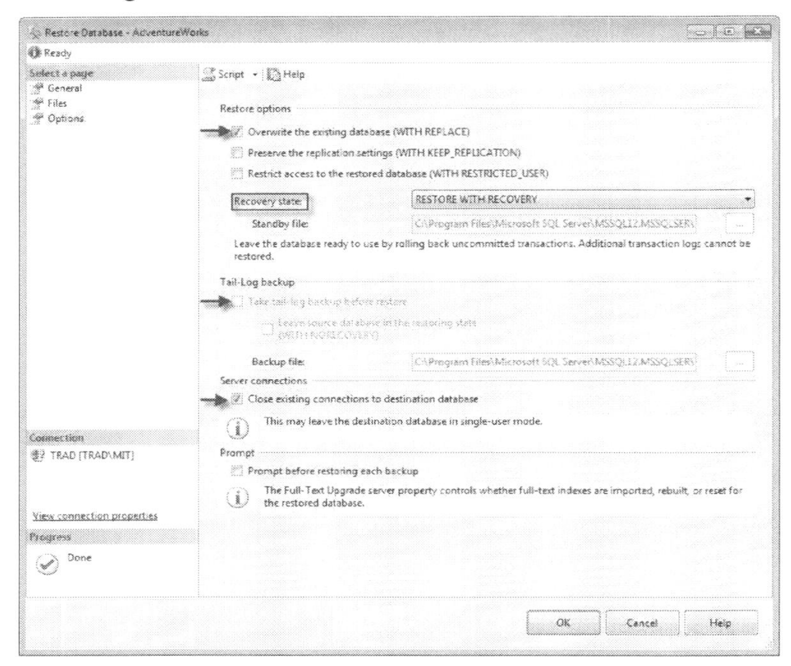

Figure 3.11 Restore Database Options Page

Next to **Recovery state** choose the option **RESTORE WITH RECOVERY** since you will not be restoring any additional files. Next if it is checked, uncheck the box to **Take tail-log backup before restore**. (This option in the restore dialog was new to SQL Server 2012. Tail-log backup will be covered in Chapter 4.)

Nobody should be connected to or using this database until you are done with the restore. You might need to tell some connections to disconnect during this restore. You can close any existing connections to the database by checking the box, **Close existing connections to destination database**. I want to point out that I typically like to check for existing connections, using SP_WHO2 system stored procedure to ensure I am not affecting any active users.

You are now ready to restore the AdventureWorks database full and differential backups. Click **OK** on the screen in Figure 3.11. A dialog box will display, confirming that the database was successfully restored (as Figure 3.12 shows).

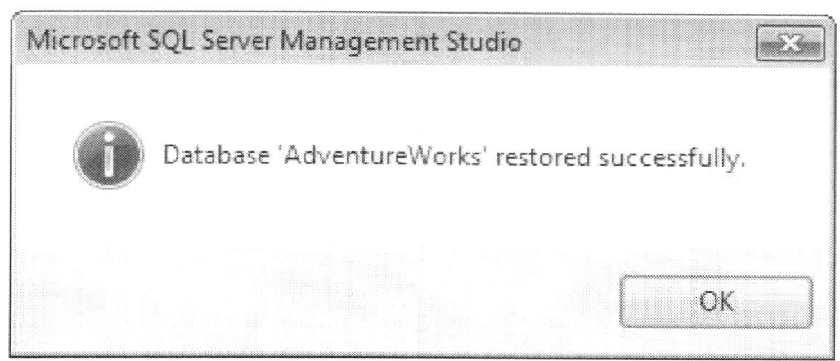

Figure 3.12 Confirmation window showing that the restore ran

Using T-SQL to Restore

As was the case in Chapter 2, the preferred choice for performing restores is to use T-SQL. Once you get your T-SQL script right, you can save the script for future use. Anytime that you need to perform another full and differential restore, you can access your saved script, make whatever changes are needed, and execute the code.

Just like you did in the graphical UI step, you have to restore the full backup that belongs to the differential database backup before you can restore the differential.

The syntax for a database restore using T-SQL is

```
RESTORE DATABASE DB_NAME FROM <backup_device> WITH
<options>
```

You can choose from several options to specify after the WITH statement. You can specify the recovery setting RECOVERY / NORECOVERY / STANDBY. In this example, you need to restore your full database with

NORECOVERY, and then you will restore your differential with RECOVERY to bring the database online.

Since the default of a restore is WITH RECOVERY, technically you do not have to specify it during the differential restore. You can also use WITH MOVE to specify the location of the database files. (You can use both WITH MOVE and RECOVERY/NORECOVERY.) You can use REPLACE to overwrite the existing database. You can use STATS to monitor the progression of the restore. Type the path and name of the database file. Replace MMDDYYYY with actual date value in the form of Month Day Year. For example, if today is June 1, 2014, then the filename for the full should be AdventureWorks_Full_06012014.BAK. This gives you a nice visual aid so that you don't have to look at the timestamp on the file itself:

```
USE [MASTER]
GO
RESTORE DATABASE [AdventureWorks]
FROM DISK = 'C:\Backups\AdventureWorks_Full_MMDDYYYY.BAK'
WITH MOVE 'AdventureWorks2012_data'
TO 'C:\Program Files\Microsoft SQL
Server\MSSQL12.MSSQLSERVER\MSSQL\DATA\AdventureWorks2012_d
ata.mdf',
MOVE 'AdventureWorks2012_log'
TO 'C:\Program Files\Microsoft SQL
Server\MSSQL12.MSSQLSERVER\MSSQL\DATA\AdventureWorks2012_l
og.ldf',
STATS = 1, REPLACE , NORECOVERY
USE [MASTER]
GO
RESTORE DATABASE [AdventureWorks]
FROM DISK = 'C:\Backups\AdventureWorks_DIFF_MMDDYYYY.BAK'
WITH MOVE 'AdventureWorks2012_data'
TO 'C:\Program Files\Microsoft SQL
Server\MSSQL12.MSSQLSERVER\MSSQL\DATA\AdventureWorks2012_d
ata.mdf',
MOVE 'AdventureWorks2012_log'
TO 'C:\Program Files\Microsoft SQL
Server\MSSQL12.MSSQLSERVER\MSSQL\DATA\AdventureWorks2012_l
og.ldf',
STATS = 1, REPLACE
```

Summary

Adding differential backups into the backup routine can significantly reduce the overall backup time and backup storage requirements. Each differential backup will back up all changed data since the last full backup. If you have a backup routine of weekly full backup and daily differentials, you will see the daily differential grow each day since each new daily backup will contain all the changes since the last full backup. Using a backup routine of full and differentials will require a change to the restore routine if you have only been using daily full backups. You will have to restore the full backup first with NORECOVERY before applying the differential backup.

Points to Ponder Differential Backup

1. A differential backup runs like a full backup but contains only the data that has changed since the last full backup.
2. Differential backups are a great mechanism to help move or migrate a database. For example, if at 10:00 PM you need to back up 500 GB of databases and restore to a new server, then at 8:00 PM you could make a full backup, copy it to the new server, and restore with NORECOVERY. Then at 10:00 PM, make a differential to capture the changed data, copy it to the new server, and then restore it.
3. Differential backups cannot be restored by themselves; you must restore the previous full (base) backup.
4. Just like full backups, differential backups can cause additional I/O, so use caution when making differential backups during peak times if there has been a considerable amount of data change since the last full backup.

Review Quiz – Chapter Three

1. A differential backup consists of which items?

a. All transactions since the last full backup
b. Only changed data since the last full backup
c. Everything needed to fully recover the database
d. Only the primary file group

2. When recovering a differential database, you have to restore the previous full database with NORECOVERY first.

a. TRUE
b. FALSE

3. When restoring the full database prior to restoring the differential backup, which recovery setting should you use to recover the full database?

a. NORECOVERY
b. STANDBY
c. ONLINE
d. RECOVERY

Answer Key

1. A differential backup contains all changed data since the last full backup and does not reset the flag/marker when backed up. Therefore answer (b) is correct.
2. To restore a differential backup, you must first restore the full backup associated with the differential. When restoring the full backup, you must leave the database in a non-recovered state so you can restore the differential. Answer (a) is correct.
3. To restore a differential backup, you must first restore the full backup associated with the differential. When restoring the full backup, you must leave the database in a non-recovered state so you can restore the differential. Answer (a) is correct.

Chapter 4. Transaction Log Backups

It is more and more common for every household to use a digital video recorder (DVR) with the TV. A benefit of having a DVR is that viewers can rewind or replay something they just watched. In a SQL Server environment, the transaction log of your database is an analogy to the DVR. The transaction log acts as a kind of recorder. As discussed in Chapter 1, the transaction log, which is part of the database file structure, is a key part of the database because it maintains the records of all database modifications.

Going back to the video recorder example: Did you know that a movie is actually 60 still pictures per second flashing in sequence to give it the look of real motion? There are 60 frames per second, so a 20-second commercial has 1,200 frames. Video editors might speak to each other in terms of which ordered frames to edit for the finished work. For example, in the 20-second commercial, the editors might say, "Let's start to fade to a blur dream effect between frames 810 and 825." The higher the frames number, the further into the video they are.

SQL Server is a motion of data instead of pictures, and it, too, marks its steps much like frames in a movie. Each data modification is uniquely identified with a log sequence number (LSN), and these LSNs are ordered. An LSN that has a higher number occurred after an LSN with a lower number. The LSNs are very important during restores because they track the order of events, as well as the point in time when data was restored.

If the transaction log held every record from the database and every change made to the database, it would be even larger than all the data files combined. This could make the database slow and storage cost too expensive. For this reason, the transaction log should be relatively small and nimble and regular maintenance of the transaction log is critical. Otherwise the transaction log will grow until the server runs out of disk space.

To maintain the transaction log and to free up the stored transactions within it, you need to regularly truncate the log. When you truncate the log you don't want to lose this data. The best way to do this is by backing up the transaction log, which is like moving the data from the transaction log into a small backup file. Not only does backing up the log help manage the size, it also decreases overall vulnerability for data loss.

Someone is likely paying you to keep their system healthy and safe. The expectations specifying how well they expect you to keep things running are called a Service Level Agreement (SLA). For example, the agreement might be that you can't allow more than 1 hour of data loss. The scheduled transaction log backups should meet or exceed the terms of the SLA. If the organization states that a system cannot sustain more than 30 minutes of potential data loss, then the scheduled transaction log backups should occur at least every 30 minutes.

Regular transaction log backups should be part of the regular backup routine for any database in full recovery model. The process of backing up and restoring transaction logs is very similar to that for differential backups. To restore a transaction log backup, you must first restore the full backup and then the differential if they are part of the backup routine. Then you can start applying the transaction logs in order.

Technically you can restore each transaction log since the last full backup. However restoring the differential backup helps drastically reduce the number of transaction log restores required.

NOTE: For the exercises in this chapter, you need to create a folder called Backups on the local C: drive (C:\Backups). You will also need a recent copy of the sample AdventureWorks database, which you can download at http://msftdbprodsamples.codeplex.com/releases/view/93587. This book is using AdventureWorks 2012, which was renamed to AdventureWorks. You will need to have run the SetupScript01.sql file. You can find all scripts mentioned in this chapter in the Book Series section at www.LinchpinPress.com.

To make transaction log backups, you must have a recovery model of the database that supports logging transactions. The two recovery models that allow this are full and bulk logged.

You can check the recovery model of the database by right-clicking on the database, choosing **Properties**, and then clicking on the **Options** page (which you see in Figure 4.1).

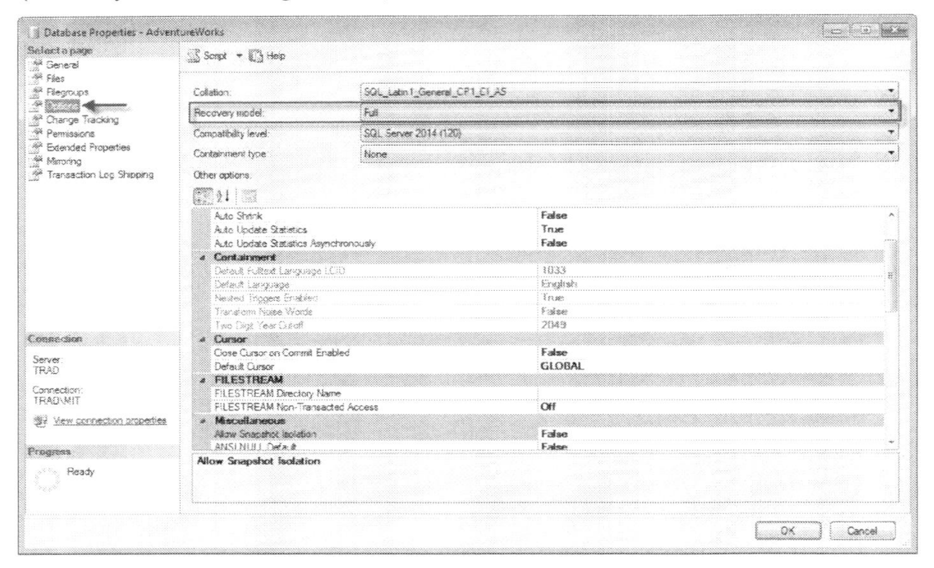

Figure 4.1 The Options page of Database Properties; view the recovery model

This example uses the AdventureWorks database, which is in the full recovery model. If your AdventureWorks database uses a recovery model that does not support transaction log backups, you can change it here to **Full** or **Bulk_logged**. If your sample database is not in Full or Bulk_logged, you can change it now.

To begin making transaction log backups after changing to **Full** or **Bulk_logged** recovery model, you need to take a full backup. Click **OK**.

At some point, you will very likely come across a database that will not allow a transaction log backup because the recovery model does not support it. In that situation, when you try to use T-SQL to back up the transaction log, you'll get an error similar to the message below:

```
Messages
Msg 4208, Level 16, State 1, Line 1
The statement BACKUP LOG is not allowed while the recovery model is
SIMPLE. Use BACKUP DATABASE or change the recovery model using ALTER
DATABASE.
Msg 3013, Level 16, State 1, Line 1
BACKUP LOG is terminating abnormally.

                                                              0 rows
```

If you are using the graphical UI to make a backup, then **Transaction Log** will not be listed. If you are trying to restore a transaction log to a database that is in an unsupported recovery model, then **Transaction Log** will not be an option, as shown in Figure 4.2.

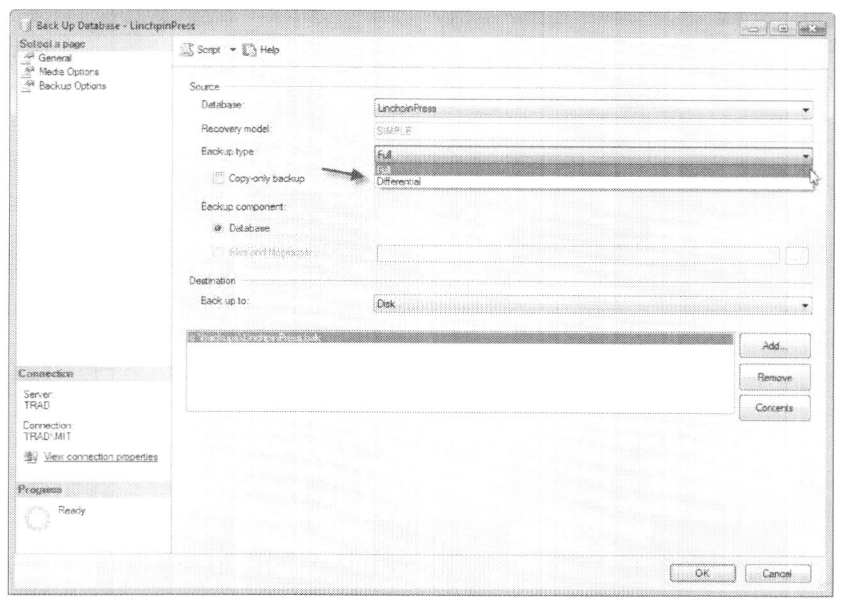

Figure 4.2 Transaction Log restore option is not listed

Since you did the steps shown in Figure 4.1, you are ready to get started on your backups. You can back up transaction logs only after a full backup, so you reuse your full backup script from Chapter 2. Execute the following script to get started:

```
BACKUP DATABASE AdventureWorks TO DISK =
'C:\Backups\AdventureWorks_FULL_MMDDYYYY_TSQL.BAK'
```

Using the Graphical UI

By now you should be very familiar with the graphical UI to make backups. To use the graphical UI within SSMS to make a transaction log backup, you need to follow a few simple steps—just like you did for the full and differential backups. First, right click on the database you want to back up, and choose **Tasks** > and then the **Back Up…** option. Do this now with AdventureWorks, as shown in Figure 4.3.

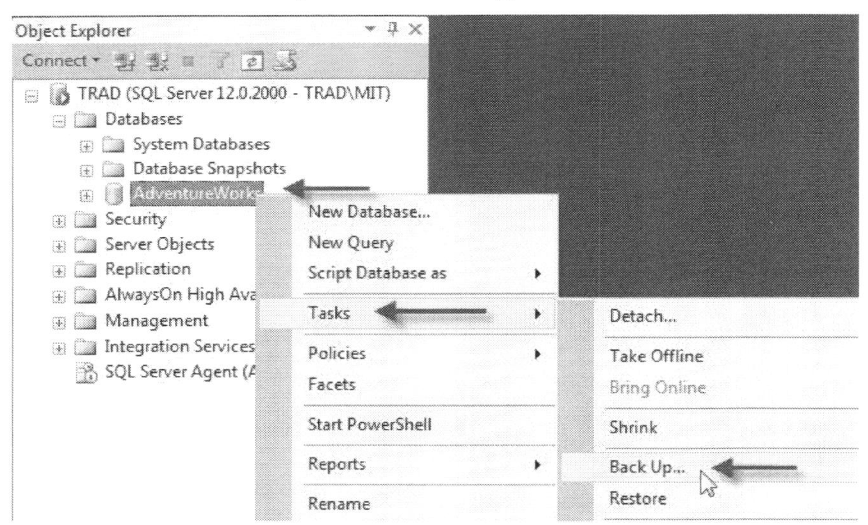

Figure 4.3 Right click on the database; choose Tasks and then Back Up…

Once you have chosen **Back Up…** , the **General** page of the **Back Up Database – AdventureWorks** box appears. Here you can choose among several selections. In Figure 4.2, the first option is to choose the database you want to back up. Next is the backup type you want to make. In this case, you want transaction log. The next item is the destination and name of the backup file. The default path is set at the instance level. For SQL

Server 2014, the default location is: C:\Program Files\Microsoft SQL Server\MSSQL12.MSSQLSERVER\MSSQL\Backup\.

For this example, you need to replace that value. So click **Remove** > **Add** (as you see in Figure 4.4).

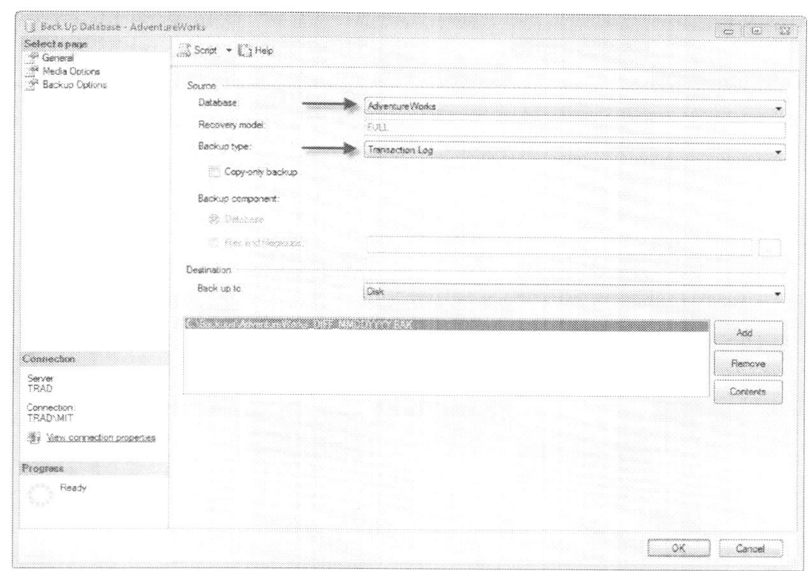

Figure 4.4 General options for backup

Type the path and name of the database backup file (as in Figure 4.5). As with full and differential backups, include the date and time, plus backup type, in the backup name. This is even more important with transaction log backups as you typically make more than one per day: All transaction log backups should have a unique file name. Type the path and name of the database file. Replace MMDDYYYY with actual date value in the form of Month Day Year. If today is June 1, 2014, 10:05.30 AM, the filename is AdventureWorks_TLOG_06012014100530.BAK. Click **OK**.

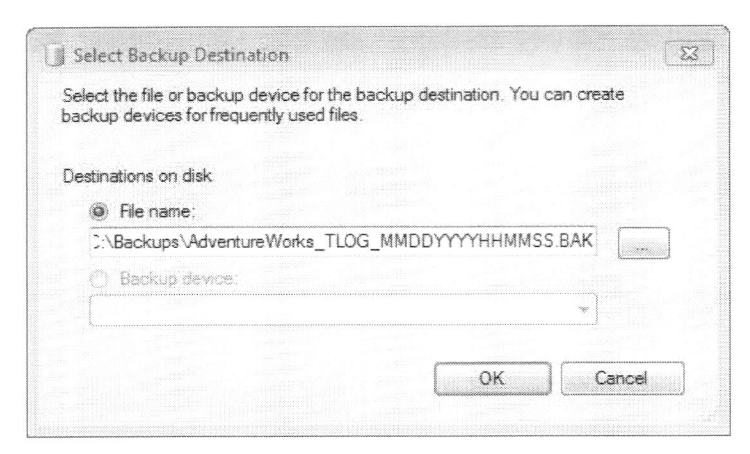

Figure 4.5 Replace MMDDYYYYHHMMSS actual values

Click the **Media Options** page for several more options. Figure 4.6 shows the default values. **Overwrite media** defaults to **Append to the existing backup set**.

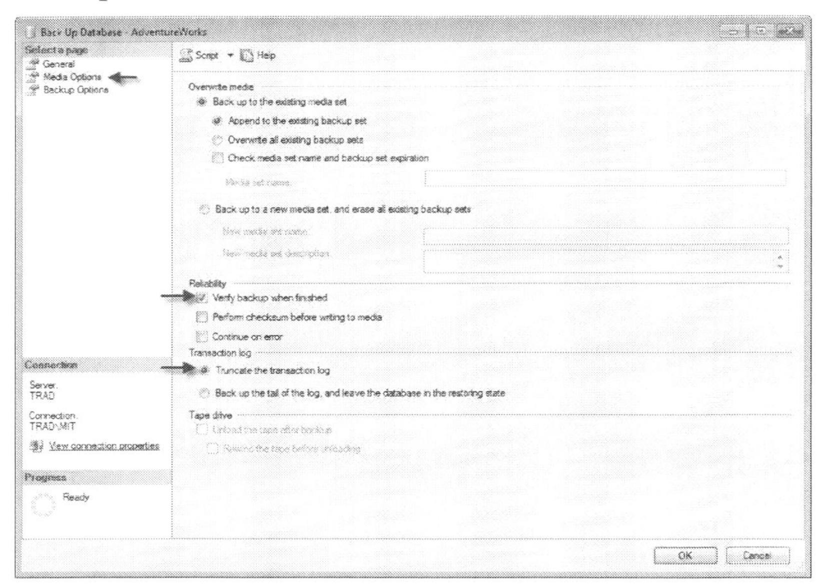

Figure 4.6 Additional options for backups

Just as with full and differential backups, you must verify transaction log backups by restoring them regularly. **Verify backup when finished** is still recommended. However, that step does not mean the backup file is 100 percent valid. If your backup solution includes making transaction log backups, then they should also be part of the restore validation checks that you perform.

By regularly restoring the full backups and applying transaction logs to them, you are testing the backup solution and getting excellent experience by performing disaster recovery tests.

If you are responsible for being able to fully recover the database environment and you are not familiar with recovering a database server up to a point in time, then practice and practice often. You can specify the media set just as you can with full and differential backups. Typically you would just accept the default value here.

In SQL Server 2012, a new section is listed for **Transaction log**. In this section you can choose to truncate the transaction log or back up the tail of the log and leave the database in the restoring state. The default here is to truncate the transaction log, which will free up log space.

The option to back up the tail log is used in preparation to restore the database and will back up the active log. If you choose this option, the database will become unavailable to all the users. Ensure that **Truncate the transaction log** is selected. Note that if the log backup fails, it's likely because your full backup did not complete first.

Click on the **Backup Options** page. Here you can specify the name of the backup set this backup belongs to, set the time the backup set will expire, as well as specify whether to use compression or not. For this example, use the default server setting for compression. Click **OK**, and the AdventureWorks database will back up to:
C:\Backups\AdventureWorks_TLOG_MMDDYYYYHHMMSS.BAK.

Using T-SQL

If using T-SQL was a benefit for making full or differential backups, it is certainly beneficial for transaction log backups, as well. Transaction log backups are made multiple times per day, in most cases several times per hour. Using T-SQL is the preferred choice for making transaction log backups.

In this example, insert a new record so that you can capture it in your transaction log. Begin by confirming that you do not have a person by the name of Muhammad Wong in your person.person table. Execute the following script:

```
SELECT LastName, FirstName
FROM AdventureWorks.Person.Person
WHERE LastName = 'Wong' AND FirstName = 'Muhammad'
```

This should get a received “(0 row(s) affected)”.

Now insert a record to add Muhammad Wong as a contractor, by executing the following script:

```
INSERT INTO AdventureWorks.Person.Person
([BusinessEntityID],[PersonType], [LastName],
[FirstName],[ModifiedDate])
VALUES
('1000','EM','Wong', 'Muhammad', '2014-06-01')
```

The syntax for backing up a transaction log is similar to other backups. Just specify

```
BACKUP LOG DB_NAME TO <backup_device> WITH <options>
```

Include _TSQL in the file name so that you do not append to the file you created using the graphical UI. Execute the following code:

```
BACKUP LOG AdventureWorks TO DISK = 'C:\Backups\
  AdventureWorks_TLOG_MMDDYYYYHHMMSS_TSQL.BAK'
```

Using the Graphical UI to Restore

Transaction log backups are only possible if you have taken a full backup first. The full backup acts as the base backup for transaction log backups. To restore transaction logs to the database, you must first restore the prior full backup that the transaction logs belong to. If you were also making differential backups, then you would first restore the full backup and then the most recent differential backup as the base for the transaction backup. After the most recent full and differential backups are restored as your base, you can restore the transaction logs (in order) since that last differential backup. If no differential is available, then you restore in order all the transaction log backups since the full backup that you restored.

To get started, you follow the steps in Chapter 2. However you will need to specify **WITH NORECOVERY** in the T-SQL script for the full backup. In the graphical UI, you specify the equivalent **RESTORE WITH NORECOVERY** on the **Options** page. Fortunately when restoring multiple files, such as full, differential, or transaction log backups, you can use the graphical UI to chain them together. Right click the database and choose **Tasks** > **Restore** > **Database** (as you see in Figure 4.7).

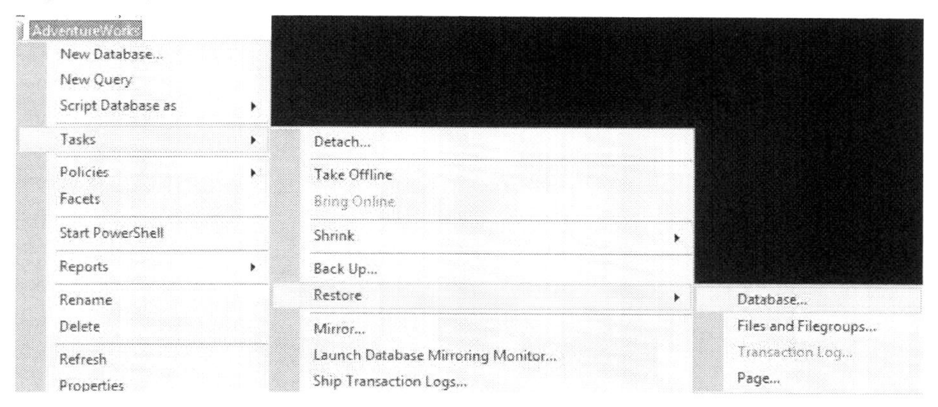

Figure 4.7 Right click on the database, chose Tasks > Restore > Database

A new window will open on the **General** page, prompting for you to choose the database source and destination. Choose **Device** as your **Source**, and then click the ellipsis (as Figure 4.8 shows).

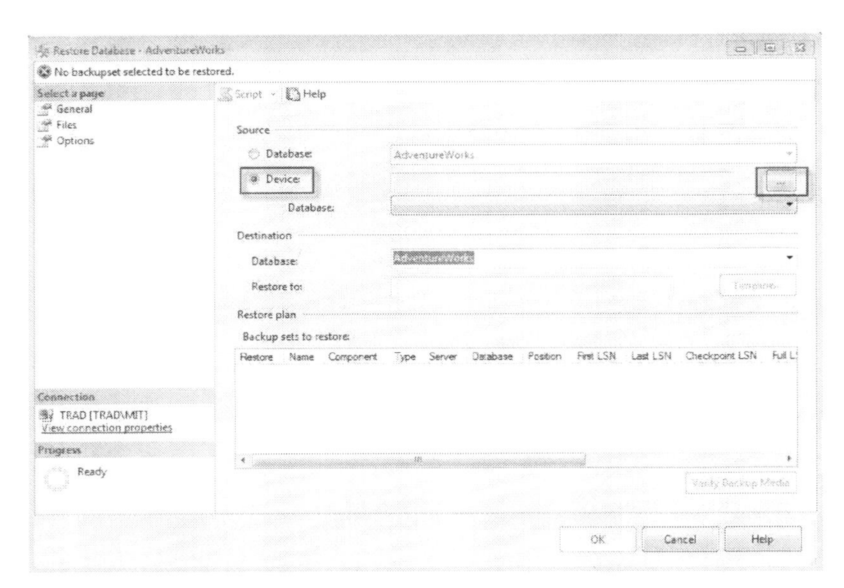

Figure 4.8 Click Device and then the ellipsis

Where are those backup files located? SQL Server will want to know this, and here is the chance to specify exactly where those files are saved. Select your backup files to restore your database, and add that exact path to SQL Server. In Figure 4.9, your **Backup media** type should be **File**, and then click **Add**.

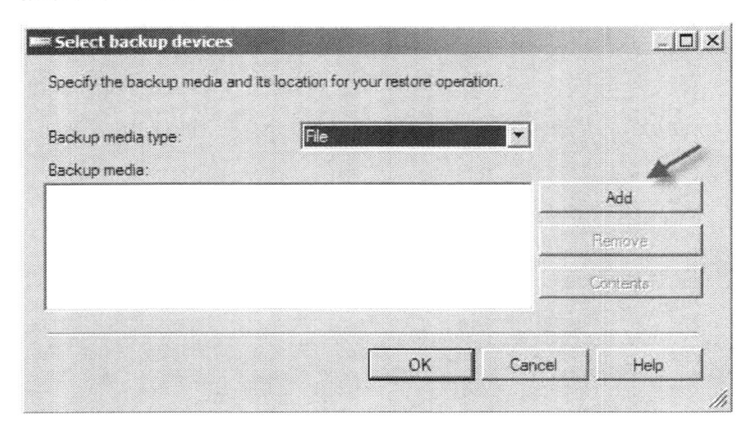

Figure 4.9 Click Add

Your files were saved to the C drive in the backup folder. For this demo, just browse to the C: \Backups folder to see the backup files you created. You can then select your full backup file and your transaction log backups that belong to the full backup. Chose **OK** and then **OK** once more (as shown in Figure 4.10).

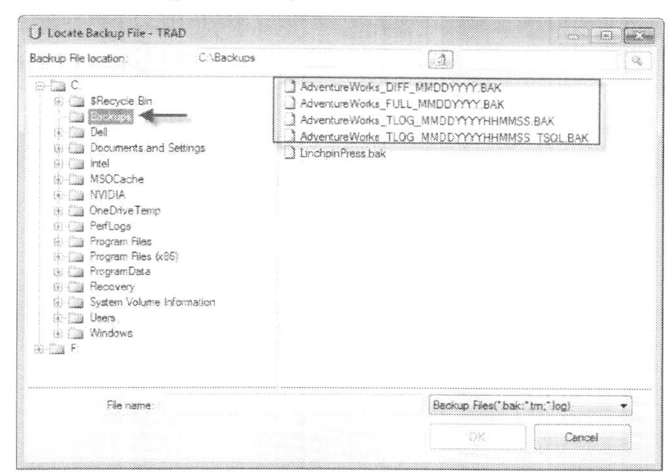

Figure 4.10 Browse to C:\Backups, select the files, and click OK

This returns you to the **Restore Database** window. Click on the **Files** page to change the location of where to restore the data and log files (as Figure 4.11 shows).

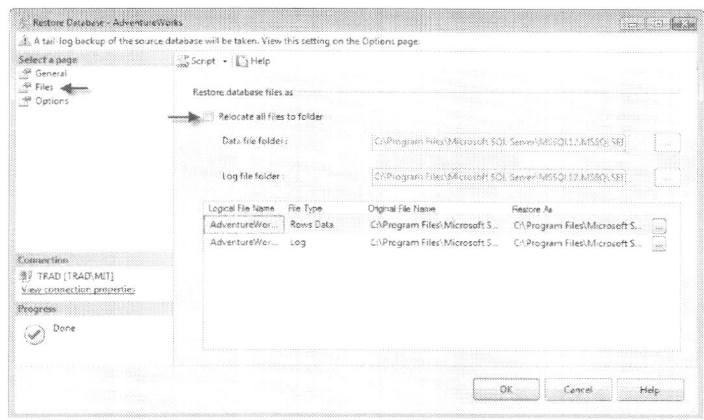

Figure 4.11 Change the location to restore the database files

You may have to restore a database to a server that doesn't have exactly the same disk configuration. Or you may want to restore the files to a location other than the default location for that instance.

Now click on the **Options** page (shown in Figure 4.12). On this page, you have a number of restore options to choose from. Check the box to **Overwrite the existing database (WITH REPLACE)**.

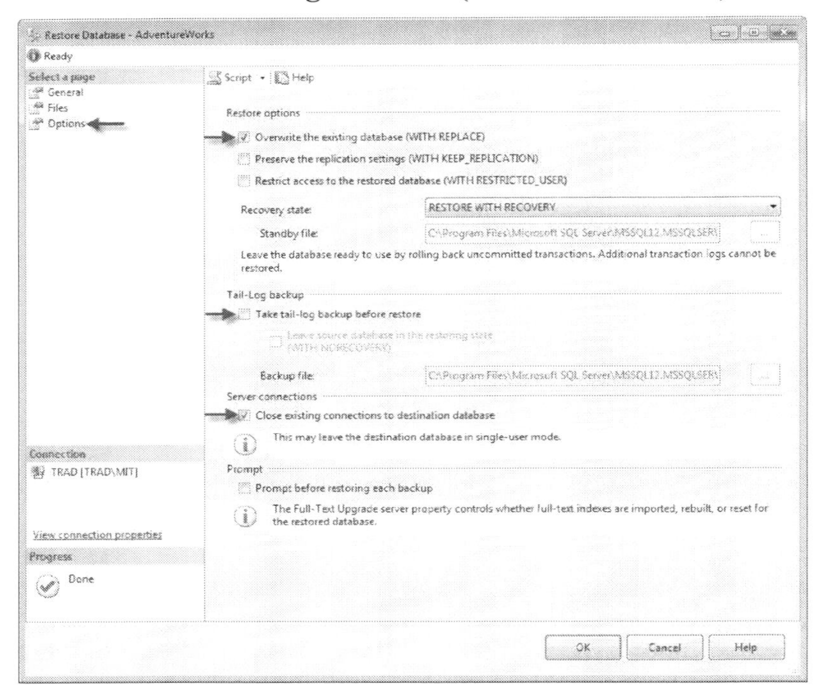

Figure 4.12 Restore Database Options Page

This is where you can leave the database offline so that additional restores can be applied, or you can bring the database online. Next to **Recovery state**, choose **RESTORE WITH RECOVERY** since you will not be restoring any additional files. Next uncheck the box to **Take tail-log backup before restore**. This option in the restore dialog was new to SQL Server 2012. (Tail-log backup terminology will be covered in Chapter 5.)

Nobody should be connected to or using this database until you are done with the restore. You might need to tell some system connections to disconnect during this restore. You can close any existing connections to the database by checking **Close existing connections to destination database**. Note that I typically like to check for existing connections, using the SP_WHO2 system stored procedure, to ensure I am not affecting any active users.

You are now ready to restore the AdventureWorks database full and transaction log backups. Click **OK** (as in Figure 4.13).

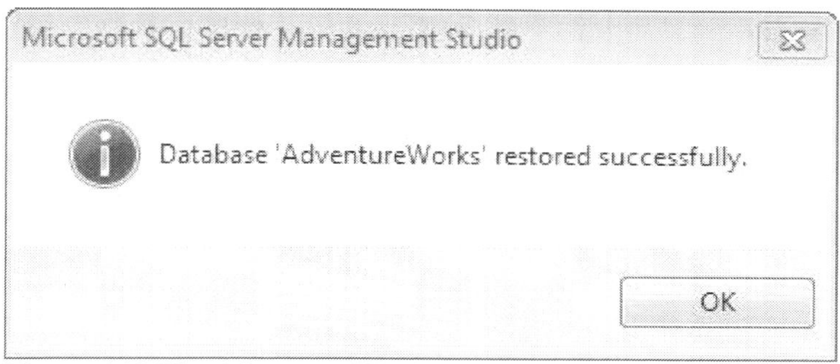

Figure 4.13 The AdventureWorks database has been restored

Using T-SQL to Restore

The preferred choice for performing restores is to use T-SQL. Although DBAs keep scripts handy for performing full and differential restores, restoring transaction logs is a little more tedious. It is a good idea to still keep a script handy, though, depending on how many transaction logs you need to restore. Scripting all of them out the first time can be a daunting task. Using T-SQL is still much easier than restoring them one by one through the graphical UI.

Just as with restoring a differential backup, you have to restore the full backup that the transaction logs belong to before you can begin restoring the transaction log or logs.

The T-SQL syntax for a transaction log restore is like this:

```
RESTORE LOG DB_NAME FROM <backup_device> WITH <options>
```

You can choose among several options after the WITH statement. You can specify the recovery setting RECOVERY / NORECOVERY. For this example, you need to restore your first transaction log with NORECOVERY, and then you will restore your last transaction log with RECOVERY.

Since the default when you restore is WITH RECOVERY, you will need to specify NORECOVERY. For this example, let's execute the T-SQL code below, substituting the ***AdventureWorks_TLOG*** backup name:

```
USE [MASTER]
GO
RESTORE DATABASE AdventureWorks
FROM DISK =
  'C:\Backups\AdventureWorks_Full_MMDDYYYY.BAK'
WITH NORECOVERY, REPLACE

RESTORE LOG AdventureWorks
FROM DISK =
  'C:\Backups\AdventureWorks_TLOG_MMDDYYYYHHMMSS.BAK'
WITH RECOVERY
```

Notice that you only restored the first transaction log that you made. This first transaction log was before you inserted the record that added Muhammad Wong as a contractor. Run your query again to show that Mr. Wong does not exist in your database:

```
SELECT LastName, FirstName
FROM AdventureWorks.Person.Person
WHERE LastName = 'Wong' AND FirstName = 'Muhammad'
```

This should get a received "(0 row(s) affected)" message. Since you have restored your AdventureWorks database **WITH RECOVERY**, you cannot restore the last transaction log.

This is a very common issue when you perform real-life restores. A risk when performing multiple restores to bring a database online is that you could accidentally restore **WITH RECOVERY** when you don't mean to. Unfortunately, you cannot set the database back into a recovery state. You must restart the restore, beginning with the full backup. You will have to restore the full backup **WITH NORECOVERY**, the second transaction log **WITH NORECOVERY**, and then your final transaction log **WITH RECOVERY**.

Give this a try now using the following script:

```
USE [MASTER]
GO

RESTORE DATABASE AdventureWorks
FROM DISK = 'C:\Backups\AdventureWorks_Full_MMDDYYYY.BAK'
WITH NORECOVERY, REPLACE
RESTORE LOG AdventureWorks
FROM DISK =
  'C:\Backups\AdventureWorks_TLOG_MMDDYYYYHHMMSS.BAK'
WITH NORECOVERY

RESTORE LOG AdventureWorks
FROM DISK =
  'C:\Backups\AdventureWorks_TLOG_MMDDYYYYHHMMSS_TSQL.BAK'
WITH RECOVERY
```

In this example, you restored both transaction logs. You added a new record after the first log backup but before the second log backup was taken. Since you just restored your full backup that does not contain your new contractor and then restored the second transaction log that contained it, you can confirm it is in fact in your database. Execute the following script:

```
SELECT LastName, FirstName
FROM AdventureWorks.Person.Person
WHERE LastName = 'Wong' AND FirstName = 'Muhammad'
```

You should now have "(1 row(s) affected)" and be able to see Muhammad Wong listed in the database.

Summary

Adding transaction log backups into the backup routine for transactional databases is probably the single most important thing you can do as a database professional. If the organization expects you to perform a point-in-time restore of an environment and you can't because you don't have adequate backups, this falls short of your SLA.

Two primary benefits of having adequate transaction log backups are keeping the size of the transaction log small and still having a point-in-time recovery capability. If you don't make regular transaction log backups on a database in the full recovery model, the transaction log will continue to grow until it runs out of drive space.

To be able to restore the transaction logs, you must first restore the last-known-good full backup and then the transaction logs in the order they were taken. Using the graphical UI to perform the restores can be time consuming as you must restore them one at a time. Using T-SQL for the restores can be much quicker to implement if they are ready ahead of time.

Points to Ponder
Transaction Log Backup

1. You can only back up transaction logs on databases using the bulk-logged or full recovery model.
2. Before you can make the first transaction log backup, you must have completed at least one full backup.
3. Each transaction log backup contains the transaction log records since the previous transaction log backup.
4. Transaction log backups issue a checkpoint, which is a way of marking a transaction as completed and officially saved to the database.
5. Data in the transaction log that has had a checkpoint is stored in two places: the transaction log and the database.
6. Transaction log backups issue a checkpoint, which allows the transaction log to free up space to be reused by moving that data to the next transaction log backup.
7. If you are restoring a database on the same server that took the backup, then the msdb will remember the order and types of your backups and make a relevant restore suggestion.
8. Because of msdb, when you use the graphical UI and you choose restore database, the most recent backup history will be displayed. On that screen, you can choose which backups to restore.

Review Quiz – Chapter Four

1. What restore sequence is correct for restoring transaction logs?

a. Full, differential, transaction logs
b. Differential, full, transaction logs
c. Full, transaction logs, differential
d. Full, incremental, transaction logs

2. Restoring multiple transaction logs on a failover server is more efficient when you use the graphical UI because it can chain multiple transaction logs and restore them at once.

a. TRUE
b. FALSE

3. Transaction logs must be restored in the sequence they were taken.

a. TRUE
b. FALSE

Answer Key

1. The restore sequence is full, differential (if you make them), and then the transaction logs in the order they were taken; therefore (a) is the correct answer.

2. On a failover server, msdb would not contain the backup history for the database you are restoring so you would have to specify one file at a time to restore them. It would be much quicker to specify the files in order in a T-SQL script to restore them; therefore (b) is the correct answer.

3. Transaction logs must be restored based on the LSN sequence, so they must be restored in order; therefore (a) is the correct answer.

Chapter 5. Common Restore Strategies

As a database administrator, you have a great responsibility to manage and maintain the data assets of your organization. To carry out this responsibility, you have to know what is expected of you. More specifically, you must know your SLAs. The type of backup and recovery approach you take can have a significant impact on how you meet your SLA. This means that you need to be aware of the various SQL Server backup and recovery options and their implications for the amount of time and resources they require and how they impact users.

With the high number of options for types of backups, you also have a multitude of options for a restore strategy. In fact. SQL Server offers more types of backups than this book has covered so far. With options such as full, differential, transaction log, file, and file group backups, you can end up with quite a few different restore strategies.

The restore strategy will depend on the backup strategy. If you have a system with data that changes daily but are only making a full backup weekly on Sunday night, you can only recover up to the most recent Sunday. You will not be able to restore to a point in time midweek.

You need to know a few things in order to create a proper backup and recovery solution. First, you must know what is expected (what your SLA is) so that you can determine the best approach.

For example, a big database will take more time to restore from a full backup than a small database. While the database is down, the people around you can get really edgy until it's back up. You need to ask the business what the Recovery Time Objective (RTO) is. RTO means: How long can your database be down?

How long does it take to restore a database? Your restore validation process can help tell you this. If you are regularly restoring your backups to validate them, you can review your restore logs to get an approximate time for recovery.

But RTO is not the only SLA point to consider. You also need to think about how current the data will be when you restore the database. This consideration is the Recovery Point Object (RPO). RPO means: How much data can you afford to lose?

If you can get the database back online in 30 minutes, some companies might consider that appropriate. But you also need to think about the RPO. Grabbing the last full backup from Sunday will take less time than grabbing the Sunday full backup *and* yesterday's differential, but taking that extra time puts back three more days of valuable data. Getting three more days of data means you get a more recent point-in-time recovery. Waiting 10 more minutes to save three days' data makes sense.

Between RTO and RPO, this book is most concerned with RPO for the backup and recovery solution. The RTO is most often a financial decision. Low RTOs usually involve building out a secondary environment for high availability, as well as an environment in a secondary datacenter. If your RPO is 15 minutes, transaction log backups every 15 minutes and shipping the logs offsite will meet this SLA. In contrast, if you have an RTO of 15 minutes, then you will have additional hardware cost for infrastructure such as a live failover system or cluster. You can control RPO by using tools within SQL Server, but RTO requires financial commitment from your company.

SQL Server has different ways it can treat the recent data stored in its log files. Most often it will save the checkpoint data until you back up the log. The log file grows until the backup is performed. Once the backup is performed and checkpoint issued, the log will be truncated and start filling up again. It's kind of like taking out the trash each week.

The default recovery model of SQL Server is to hold on to the data. If data were not saved in the transaction log, then there would be no log to back up. The only choices would be full or differential backups. If recovery is not an issue and the goal is to keep the log file small, you can set the transaction log to truncate all data on the checkpoint. This is a fancy way of saying the transaction log will not hold on to any data that it is not currently being worked on. No storage of data. Truncating the log on checkpoint happens when you set the recovery model to simple mode.

Depending on the backup needs, each database will need to have the correct recovery model. If the database is in simple mode, then you will not be able to make any transaction log backups. You will still be able to schedule full and differential backups while using the simple recovery model. For some companies with a relaxed RPO, this may be an adequate solution.

No one solution fits all situations for backup and recovery. Often the safest solution is one that also includes regular transaction log backups. However, in some situations this could be overkill and use up valuable resources unnecessarily.

The best thing you can do to determine the best approach to backup and recovery is to have the business define its needs. Then you can architect a solution to meet and exceed the requirements.

To help you understand the options and when to use them, let's now explore some possible scenarios and strategies. The following sections will list out various scenarios and define the RPO they provide.

> **NOTE:** For the exercises in this chapter, you need to create a folder called Backups on the local C: drive (C:\Backups). You will also need a recent copy of the sample AdventureWorks database, which you can download at http://msftdbprodsamples.codeplex.com/releases/view/93587. This book uses AdventureWorks 2012, which is renamed to AdventureWorks. You will need to have run the SetupScript01.sql file. You can find all scripts mentioned in this chapter in the Book Series section at www.LinchpinPress.com.

Full Backup Only

If you can either recreate daily work that might be lost, or if no data changes throughout the day, a full backup could be the only type of backup you use. Full backup is an adequate solution for those two scenarios. However, if the database has some form of data that changes throughout the day and it cannot be easily recreated in case of a failure, then you need a more robust backup solution than just full backup. In this case, your options are full backup with differential backups and full backup with transaction log backups.

Full Backup with Differential Backups

Full backups with differential backups can do a couple of things for you. First, if you're performing weekly full backups with daily differentials, you are decreasing the amount of data that has to be re-entered on a daily basis. For an RPO, this approach does not give you any increased benefit, except for saving storage space as compared to taking daily full backups.

If you're making multiple differential backups per day, then this decreases the RPO by the amount of the time difference of the differential backups. This approach would be beneficial if you had a system that only imported data every 4 hours and the database were scheduled to be backed up every 4 hours. If you performed nightly full backups and scheduled a differential every 4 hours (one for each 4-hour data load), then you would be reasonably protected.

Full Backup with Transaction Log Backups

The most common backup and recovery solution for Online Transaction Processing (OLTP) databases is nightly full backups with transaction log backups at a short interval. Common time frames are 1 hour, 30 minutes, 15 minutes, 10 minutes, and 5 minutes.

The more frequently the transaction log backups occur, the more transaction logs you will be dealing with during a restore. If you've been following along in the book and participating in the exercises in Chapter 4, you know that restoring multiple transaction logs could be a very time-consuming task. Restoring transaction logs takes time using either the graphical UI or T-SQL (if you had to write out each and every transaction log to restore).

There is a much simpler way to create the restore scripts than having to manually recreate them or than using the UI. Msdb stores all backup history for every database. The two tables to take a look at are MSDB.dbo.backupset and MSDB.dbo.backupmediafamily. You can join these two tables together and generate a robust script that will get the last full backup and each transaction log backup since the most recent full.

In the backupset table, the **type** field specifies the type of backup. Full = D, Differential = I and transaction = L. To show a small result set that demonstrates what this data looks like, run the following script:

```
SELECT TOP 10
  b.type,
  b.database_name,
  mf.physical_device_name
FROM msdb.dbo.backupset b
  JOIN msdb.dbo.backupmediafamily mf
ON b.media_set_id = mf.media_set_id
ORDER BY backup_set_id DESC
```

You will get a similar result set to what you see in Figure 5.1. The result set below shows that the backup set **type** for full is D, differential is I, and transaction log is L.

	type	database_name	physical_device_name
1	L	AdventureWorks	C:\Backups\AdventureWorks_TLOG_MMDDYYYYHHMMSS_TS...
2	L	AdventureWorks	C:\Backups\AdventureWorks_TLOG_MMDDYYYYHHMMSS.BAK
3	I	AdventureWorks	C:\Backups\AdventureWorks_DIFF_MMDDYYYY.BAK
4	D	AdventureWorks	C:\Backups\AdventureWorks_Full_MMDDYYYY.BAK

Figure 5.1 Sampling of data

With some basic T-SQL skills, you can generate a query to create the restore scripts. An included sample script is in the resources folder named **(**RESTORE_SCRIPT_FULL_W_TLOG.SQL**)**. When you execute this script for the database you need to restore, the script will query msdb and get the most recent full backup and each transaction log backup since the most recent full backup was taken. Each file restore uses WITH NORECOVERY with a final script that restores the database WITH

RECOVERY to bring it online. Figure 5.2 shows a sample of the result set.

Figure 5.2 Result set sample for full backup with transaction log

backup_set_id	Script
2014	RESTORE DATABASE AdventureWorks FROM DISK = 'C:\Backups\AdventureWorks_Full_MMDDYYYY.BAK' WITH NORECOVERY
2016	RESTORE LOG AdventureWorks FROM DISK = 'C:\Backups\AdventureWorks_TLOG_MMDDYYYYHHMMSS.BAK' WITH NORECOVERY
2017	RESTORE LOG AdventureWorks FROM DISK = 'C:\Backups\AdventureWorks_TLOG_MMDDYYYYHHMMSS_TSQL.BAK' WITH NORECOVERY
2147483647	RESTORE DATABASE AdventureWorks WITH RECOVERY

All you need to do is copy the script column and paste it into your query window and then execute the script.

Imagine this is a production system that is making transaction log backups every 5 minutes. In 10 hours, that would be 120 transaction log backups. Writing the restore script would take a while and is certainly not something you want to have to do while customers and your boss are waiting for you to recover a critical system.

I recommend that you add a step to each of your scheduled backup jobs that will create the restore script as you make the backups. Even if the directory structure is different on the server you are performing the restore on, you can easily find and replace to update drive letters and paths. Having a script per database that writes to a directory that is backed up with your database backups can drastically decrease the complexity and the time it takes to begin the restore process.

Full, Differential, Transaction Log

Using a nightly full backup and daily differential and regularly scheduled transaction log backups is a very common backup solution. It provides the same level of protection as nightly full backup and transaction log backups, however it helps to reduce the daily backup time by only backing up the changed data. As Chapter 3 covered, having differential backups in the backup routine means you have to restore the full backup, then restore the latest differential, and then any transaction logs since that differential. It is also possible to restore the full backup, then all the transaction logs since the full backup. This would be much more time consuming than just restoring the full, most recent differential, and then the remaining logs.

The same concept applies here as with nightly full and transaction log backups. Restoring the full backup and then the differential is the easy task; applying the transaction log backups is time-consuming (since there are often many transaction logs and they need to be done in order). The same process for using msdb to gather the last full, differential, and then all transaction logs since the most recent differential applies here.

A sample script in the resources folder called RESTORE_SCRIPT_FULL_DIFF_W_TLOG.SQL. When you execute this script for the database you want to restore, it will query msdb for that database. It gets the most recent full backup, the most recent differential since the last full, and each transaction log backup since the last differential was taken. Each restore uses the WITH NORECOVERY until the final script. The final restore uses the WITH RECOVERY. You can see an example of the result set in Figure 5.3.

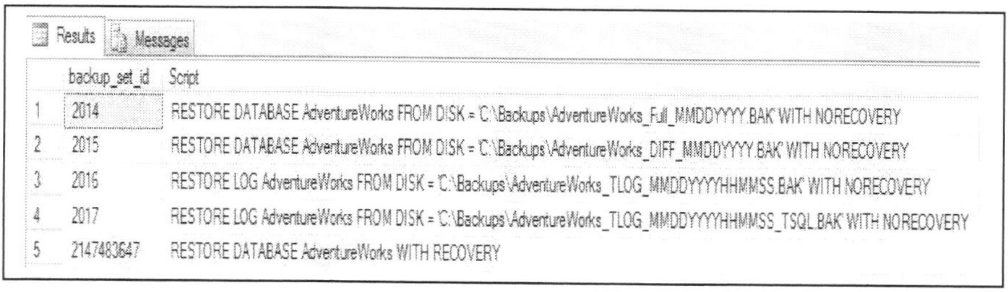

Results | Messages

	backup_set_id	Script
1	2014	RESTORE DATABASE AdventureWorks FROM DISK = 'C:\Backups\AdventureWorks_Full_MMDDYYYY.BAK' WITH NORECOVERY
2	2015	RESTORE DATABASE AdventureWorks FROM DISK = 'C:\Backups\AdventureWorks_DIFF_MMDDYYYY.BAK' WITH NORECOVERY
3	2016	RESTORE LOG AdventureWorks FROM DISK = 'C:\Backups\AdventureWorks_TLOG_MMDDYYYYHHMMSS.BAK' WITH NORECOVERY
4	2017	RESTORE LOG AdventureWorks FROM DISK = 'C:\Backups\AdventureWorks_TLOG_MMDDYYYYHHMMSS_TSQL.BAK' WITH NORECOVERY
5	2147483647	RESTORE DATABASE AdventureWorks WITH RECOVERY

Figure 5.3 Result set sample for full, differential, and transaction logs

Having a step in your scheduled backup jobs to script out the restore scripts to a file is key to being able to quickly have the restore scripts ready.

Summary

As a database administrator your primary responsibility is likely to be protecting your organization's data. That means that you need to be able to recover the database in the event of a disaster or data quality issue.

If you take only full backups, your restore strategy is to use the most recent full backup. If you are taking full and differential backups, you need the most recent full and the most recent differential. If you are taking full, differential, and log backups, you need to use the most recent full and the most recent differential followed by every transaction log backup in order after the last differential.

The SLA for the system determines the type of backup and recovery solution needed. In other words, do what the company needs. When you determine the appropriate backup solution, the key objective is to be able to meet or exceed the RPO.

The question to ask the business unit is, "How much data can you afford to lose?" The first answer you will usually get is that no data loss is acceptable.

You will have to kindly present the challenges and hardware costs that come with zero data loss. Live clustering failover systems with near zero data loss often triple the hardware costs for your company. Try to explain how reasonable and low-cost it is to allow for 15 minutes or less of data loss. In many cases, that is acceptable. In situations where that is not tolerable, you might be able to just make adjustments such as decreasing the time between transaction log backups from 15 minutes to 5.

Organizations are spending much more time and money to ensure that their critical systems are highly available. SQL Server includes several technologies that you can apply to mitigate the risk of extended downtime. Mitigating extended downtime is referred to as high availability (HA). HA is not disaster recovery. RTO dictates how quickly the database would need to be recovered or roughly translated as, "How long can the system be down?" Typically systems that have an established RTO have a dedicated system set up for disaster recovery.

When you think of disaster recovery solutions, the idea of database mirroring will arise, but be aware that mirroring will not satisfy the RTO need for a disaster recovery system. Database mirroring involves having a dedicated server in place that provides a hot standby on a database-by-database basis. Many people incorrectly think asynchronous database mirroring is a disaster recovery solution. Database mirroring unfortunately is not suitable for disaster recovery. Think about a scenario where a table was accidently dropped. In a mirror setup, the table is also dropped on the secondary server. To recover from that situation, a restore from a backup is the only option.

RTO is a measurement of how long it takes to notice an issue, report the issue, notify the person to correct the issue, and include the response time of the person needs to correct the issue. The response time of the person making the correction may involve travel time to the office or home if he or she is out to dinner and on call, the time to remote in, time to assess the situation, and the time it takes to perform the necessary actions to recover. If the restore process takes an hour, how many extra minutes are involved in the notification process, travel time, etc.? You have to factor this additional time into the recovery solutions.

Whichever backup and recovery plan you put in place, you need to test it and test it often. You don't want to be in a situation where you are recovering a critical system and the CXO asks how long it is going to take to restore if you don't even know yourself. Wouldn't it be much nicer to pull out a log and tell the CXO that the last time this was performed (within a month or so) that it took X number of minutes?

When documenting your disaster recovery plan, write up your restore process in enough detail that anyone who can work a keyboard can follow your procedures to recover the databases. If a disaster strikes that consumes your datacenter, it is very likely that any IT staff your company has (you included) within 200 miles of your datacenter would also be affected by the event. Your attention would be on your own safety, as well as that of your family, rather than on your company's assets. Plan for someone in another state to be able to execute your restore strategy. Plan for the worst, and hope for the best. Having those restore scripts scripted out is a great way to enable someone else to bring your databases back online.

Points to Ponder
Common Restore Strategies

1. When you have multiple files to restore, each restore uses the WITH NORECOVERY until the final statement. The final restore uses the WITH RECOVERY.
2. If you had 10 backup files to restore (1 full, 1 differential, and 8 transaction log backups), then the full, differential and the first 7 transaction log backups would be NORECOVERY and the last transaction log backup would be RECOVERY.
3. Backups are only valid if they can be restored. Without a regular process to validate your backups, you are more likely to suffer extended downtime and potential data loss.
4. It is entirely possible to recover a database from an initial full backup and every transaction log since the full was taken. In scenarios with a weekly full backup, daily differential and transaction log backups every 10 minutes, you can restore the full backup and each transaction log backup, skipping the differentials. It takes much more time and is more complex, but if you do not have the differentials it is possible.
5. Whichever backup strategy you use, test it fully to ensure it meets the SLA of the organization. Being able to restore a database within the limits of the SLA could mean the difference in the company staying open for business or closing its doors.
6. In the backupset table, the **type** field lists the type of backup. Full = D, Differential = I and Transaction = L.
7. If a database is 500 GB and the RTO is 60 minutes, chances are you can't be notified, connect to the remote server, and restore from backup within 60 minutes. If you configure log shipping with a 24-hour load delay, you can easily roll the logs forward, up to seconds before the issue that is causing you to have to restore. The transaction logs from the 24-hour period of time are far less than the 500 GB full backup plus the logs since the backup was taken.

Review Quiz – Chapter Five

1. RPO stands for ___________________?
   a. Recovery Process Owner
   b. Recovery Product Object
   c. Recovered Process Online
   d. Recovery Point Objective
2. A restore script can be created by using the data stored in msdb.
   a. TRUE
   b. FALSE
3. For a high transactional database, which option gives the most protection for recovering data with the least amount of data loss?
   a. Weekly full backups, daily differential with transaction log backups every 15 minutes
   b. Nightly full backups
   c. Weekly full backups, daily differential
   d. Nightly full, differential backups every 4 hours

Answer Key

1. RPO = Recovery Point Objective, so (d) is the answer.
2. Msdb records every backup operation, so the answer is (a) TRUE.
3. Weekly full backups with daily differential and transaction log backups every 15 minutes will only allow for up to 15 minutes of data loss, so (a) is the correct answer.

Chapter 6. Copy Only Backup

When I think about SQL Server copy-only backup, it takes me back to the 1980s. Anyone who was a teen in the 80s is fully aware of what it is like to make a copy. Teens used to make copies of cassette tapes of their favorite music. Making mixed tapes was a really big thing to do. When compact discs came out, it was a little harder to make a mixed CD but soon people figured it out. Photo copy machines were also common. When people had to go to a library to do research, they could make a photocopy of the few pages they needed and go back to their own place and do their term papers.

The reason all this copying reminds me of SQL Server is that copying pages in a book and favorite pieces of music did not alter the original items. Similarly, if I want to copy data without changing it, SQL Server can make a backup without affecting the original data.

SQL Server 2005 and above include an option to make a backup using WITH COPY_ONLY. As you learned in previous chapters, when SQL Server makes a backup, it uses an LSN to track these backups. So what is a copy backup? This COPY_ONLY backup will back up all the data in your database without resetting any of the differential bitmaps (i.e., the backup markers, or flags) or altering the size or flow of the existing transaction log. Because those changes do not occur, the copy backup has no effect on the sequence of backups of the next differential or log file backup you do during the week. I like to tell people that it is a backup that never happened.

Why is a COPY_ONLY backup important? COPY ONLY backups help you maintain the restore chain when you need a copy of the data outside your normal backup schedule. If you do a full backup to get that copy, that backup can interfere with planned differential backups. To prevent problems, you can simply use COPY_ONLY to back up the data without changing differential bitmaps.

Let's look at a situation to illustrate why COPY_ONLY is necessary. If your organization is doing differential backups, those backups are dependent on the previous full backup.

Assume your organization is using weekly full, daily differential, and hourly transaction log backups. Now imagine that a member of your team is asked to make a current full backup of a production database to be used in dev or QA on a Tuesday morning.

Your team member (being the nice person he or she is) makes a full backup on Tuesday night and restores it to the QA server. Later in the week on Thursday, a problem with the production database occurs and the team has to restore from backup.

You are the person that has to perform the restore, so you grab Sunday night's full and Wednesday night's differential backups, right? When you try to apply Wednesday's differential backup, you get a nasty error telling you that the differential you're using does not apply to that full backup you just restored. A full backup was taken on Tuesday and nobody else knew about it—until now.

After asking the team, you find out that another backup has been taken since Sunday, but the person who made the backup has deleted it in the meantime. What are the options now?

You might not be out of luck with being able to recover the database, but it will take a lot more work to get it up and running. If you are fortunate enough to have kept enough of the transaction logs, you can apply all the transaction logs since the last full backup. That means you have to do about 100 restores: one incremental per hour since Sunday.

The scripts that you learned in Chapter 5 would be helpful in this situation. Manually creating a script of four days' worth of transaction logs can be a time consuming task.

What type of backup should have been done on Tuesday night? The answer is WITH COPY_ONLY. It's what you use whenever you make a backup outside of your regularly scheduled backup jobs.

You can prevent this type of situation when you need a complete backup of the database at an ad-hoc time by always using WITH COPY_ONLY. COPY_ONLY also applies to transaction log backups. COPY_ONLY backups for transaction logs do not truncate the log, which is very important.

Be aware that if you use COPY_ONLY with DIFFERENTIAL, the COPY_ONLY is ignored and a differential backup is taken.

NOTE: For the exercises in this chapter, you need to create a folder called Backups on the local C: drive (C:\Backups). You will also need a recent copy of the sample AdventureWorks database, which you can download at http://msftdbprodsamples.codeplex.com/releases/view/93587. In this book we are using AdventureWorks 2012, which we renamed to AdventureWorks. You will need to have run the SetupScript01.sql file. You can find all scripts mentioned in this chapter in the Book Series section at www.LinchpinPress.com.

Using the Graphical UI

To make a COPY_ONLY backup using the SSMS graphical UI, you need to follow a few simple steps, just like you did in Chapter 2. First, right click on the database and choose **Tasks** >. Select the **Back Up…** option. Do this now with AdventureWorks (as Figure 6.1 shows).

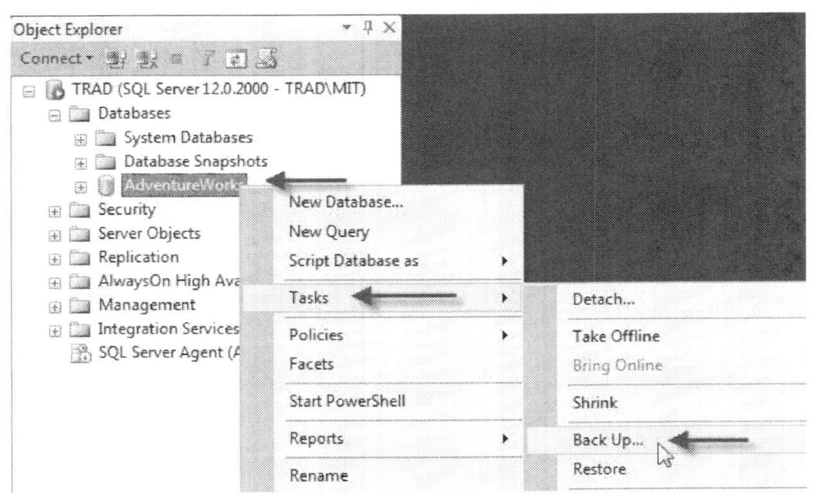

Figure 6.1 Right click on the database, choose Tasks and then Back Up...

Once you have chosen **Back Up…** , the general dialog box appears. Here you can select from several options. Selecting **OK** at this point creates a full backup of the database into the default backup location for the SQL Server instance.

In Figure 6.2, you see that the first option is to choose which database to back up. Next is the type of backup you want to make. In this case, choose **Full**. Check the option for **Copy-only Backup**. For **Backup component,** make sure **Database** is selected. The next item is the destination and name of the backup file.

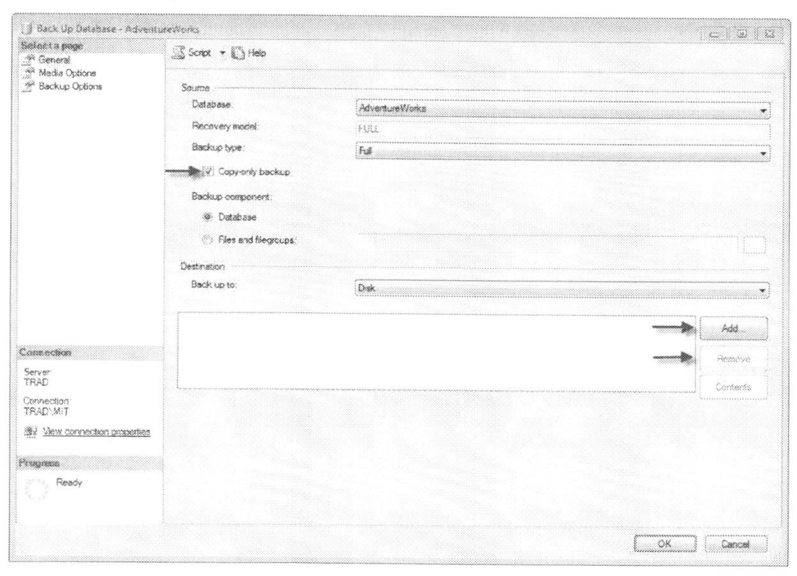

Figure 6.2 Click Copy-only Backup and remove the default destination path; add your path

The default path is set at the instance level. For SQL Server 2014, the default location is the following path:

C:\Program Files\Microsoft SQL Server\MSSQL12.MSSQLSERVER\MSSQL\Backup\

In this example, you do not want to save in that location. To change this for the example, select the path to highlight it and then click **Remove** >. Then click **Add** to make a new path.

To pick your own location, enter the path and name of the database backup file to be created (as shown in Figure 6.3).

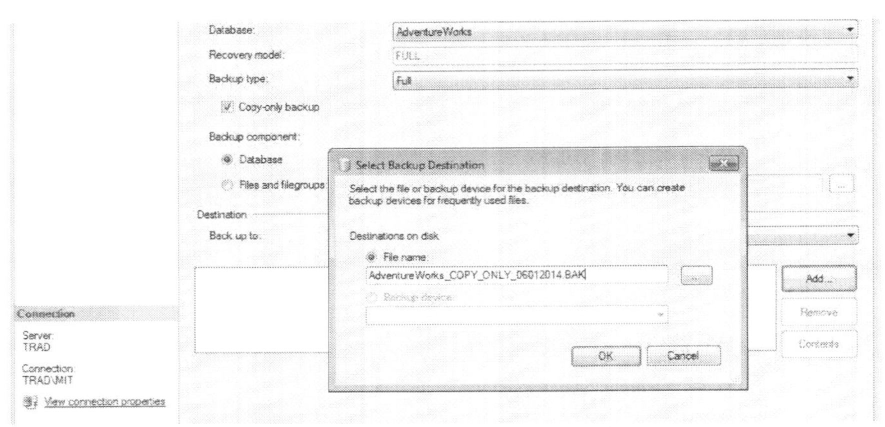

Figure 6.3 Type the path and name of the database file

A best practice is to include a date and time, as well as backup type, in the backup name. In this example, change the backup file name to AdventureWorks_COPY_ONLY_MMDDYYYY.BAK. Type the path and name of the database file. Replace MMDDYYYY with actual date value in the form of Month Day Year. For example, if today is June 1, 2014, then the filename should be AdventureWorks_COPY_ONLY_06012014.BAK. This gives you a nice visual aid so you don't have to look at the timestamp on the file itself. Click **OK**.

The choices so far in the **General** page of the **Back Up Database** dialog are just a few of the options available. Just below **General**, click **Media Options** to see several more possible backup choices. Figure 6.4 shows the default values that are already checked.

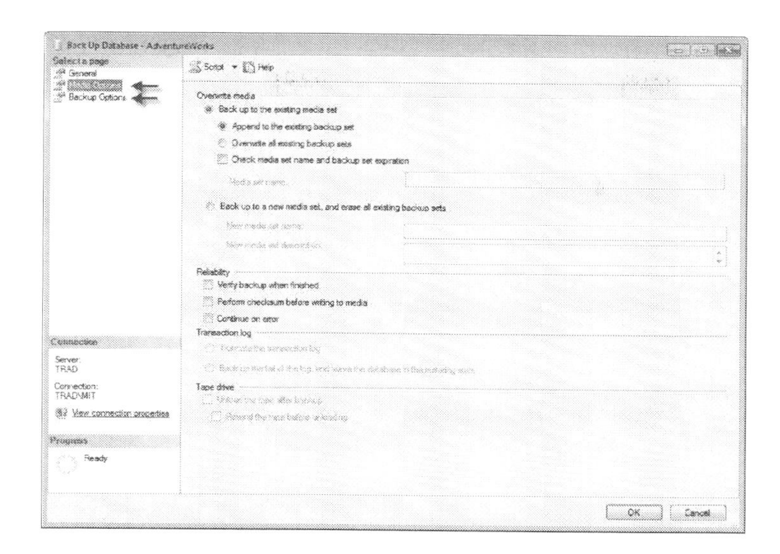

Figure 6.4 Additional options for backups.

Just as I do with full backups, I recommend that you select **Verify backup when finished**. However, that verification does not mean the backup file is 100 percent valid. In most cases, the reason for making a COPY_ONLY backup is to immediately restore it to another server. COPY_ONLY backups are typically not part of a scheduled backup plan.

Compression is also an option with COPY_ONLY backups, and it performs the same way as it does with other backups. You will use the default option here as you did in the previous chapters. Click **OK**, and the AdventureWorks database will back up to:
C:\Backups\AdventureWorks_COPY_ONLY_MMDDYYYY.BAK.

Using T-SQL

Chapter 2 introduced the benefits of using T-SQL and explained how having to check so many boxes each time you do a backup can be cumbersome. Creating a COPY_ONLY backup is no different. The syntax is similar to the syntax used in Chapter 2, except you add the

COPY_ONLY option and name your backup file to reflect the backup type. Being able to save this script for reuse at a later time makes it the preferred choice for making COPY_ONLY backups.

The syntax is straightforward. You specify:

```
BACKUP DATABASE DB_NAME TO <backup_device> WITH <options>
```

Then you specify your options. Since you are making a COPY_ONLY backup, you have to specify WITH COPY_ONLY. To make sure this file has a different name than in the file in the UI example, mark this backup file name to include _TSQL in the file name. This way, you have the backup files you used in the UI, as well as the new one you are creating now.

Type the path and name of the database file. Replace MMDDYYYY with actual date value in the form of Month Day Year. For example, if the backup was made on June 1, 2014, then the filename could be C:\Backups\AdventureWorks_COPY_ONLY_06012014.BAK:

```
BACKUP DATABASE AdventureWorks TO DISK =
  'C:\Backups\AdventureWorks_COPY_ONLY_MMDDYYYY_TSQL.BAK'
WITH COPY_ONLY
```

Using the Graphical UI to Restore

Using the graphical UI to restore a copy only backup is no different than restoring a full backup. Since the backup otherwise does not behave any differently, the restore process is the same as the full restore you performed in Chapter 2.

Summary

Using COPY_ONLY backups helps maintain the restore chain. They should be mandatory for one-off backups in organizations that use differential backups. To make COPY_ONLY backups, the process is the same as making a full backup, except you use the WITH COPY_ONLY syntax.

Points to Ponder Copy Backup

1. If you want to take an ad-hoc full backup of your database and not have it reset the differential bitmaps, you can use the WITH COPY_ONLY option on the backup.
2. COPY ONLY backups should be a standard option if you make backups outside of your regular backup cycle.
3. Backups do not cause blocking, contrary to any myths out there. However backups are very I/O intensive, so they can cause performance issues related to I/O if you run them during peak times.
4. You can use COPY_ONLY backups with all recovery models.
5. You cannot use a COPY_ONLY backup as a differential backup.
6. Using the COPY_ONLY option on backups of transaction logs does not truncate the log. If you use COPY_ONLY on a differential backup, SQL Server ignores the COPY_ONLY option and a differential backup is taken.
7. Restoring a backup that was made with the COPY_ONLY option is not any different than restoring the same type of backup without the COPY_ONLY option.

Review Quiz – Chapter Six

1. A COPY_ONLY backup of a transaction log truncates the log.
   a. TRUE
   b. FALSE
2. You can use the COPY_ONLY option to make a COPY_ONLY differential backup.
   a. TRUE
   b. FALSE

3. COPY_ONLY backups do not impact the sequence of backups.

   a. TRUE
   b. FALSE

Answer Key

1. Transaction log backups with the option COPY_ONLY do not truncate the log; therefore (b) is the correct answer.

2. Using COPY_ONLY with a differential backup causes the COPY_ONLY option to be ignored and a differential backup is taken; therefore (b) is the correct answer.

3. COPY_ONLY backups do not impact the sequence of backups. That is their unique feature; therefore (a) is the correct answer.

Chapter 7. File and Filegroup

Organization is a part of your day-to-day life. Without structure and organization life, would be chaos. Even in the simplest forms, organization helps you to be more efficient. Think about where you store your clothes. Most people have a drawer for their undergarments, a separate drawer for socks, maybe one for undershirts, one for shorts, and so forth. These drawers may be messy, but certain items are always stored in their specific place. How chaotic would your mornings be if you had to search multiple drawers for a pair of socks, only to find one color sock in one drawer and the matching sock in another?

For the sake of organization in SQL Server, you often need to split your very large databases (VLDBs), into multiple file groups. How large a database has to be in order to be considered a VLDB is debatable, but a general consensus is that anything nearing or exceeding 1 TB can be called a VLDB.

A big issue with VLDBs is the time it takes to back them up each night. Frequently, the backup does not fit within the standard backup window, and the backup runs into production hours. This overlap can impact production performance due to the I/O strain the VLDB backup can put on the disk and network. If daily backups take more than 24 hours to run, then this is a sign that you need another solution.

At some point in your career as a database administrator, you will most likely find yourself having to manage a VLDB. Either you will have to tame one by possibly splitting it into multiple file groups using the existing database design, or you may have to partition very large tables.

Before jumping straight into trying to split the database into file groups, first check the data being stored to make sure you are keeping only the data within the appropriate retention period. How old is the oldest data in the database? Often companies do not purge their data regularly. For these systems, a purge may be all that is needed to get VLDBs back to a manageable size.

When this is not the case, forecast how large the database is expected to grow and then work with the DBAs to come up with a long-term plan.

Often this involves partitioning the database. By partitioning the data into multiple filegroups you can better manage the data from index optimizations, backups, restores, and performance improvements.

In the cases I have encountered, either the design of the database already accounted for storing the data in quarterly tables, or only a couple of tables occupied the majority of the space in the database. In the case of the database with quarterly tables, I was able to easily move the tables into yearly filegroups. With the databases with very large tables with date/time columns, I used a partition function to split the data into yearly file groups. Note that these activities should always be thoroughly tested and approved by the vendors whose databases you are working with.

An additional benefit of multiple filegroups is that you can place the filegroups on separate disk arrays to improve the overall disk I/O. This also lets you place archive data on a slower disk and save the organization money by moving less frequently read data onto a cheaper and slower disk. (To give you an idea of what this can mean in terms of cost, you can figure that high-performance enterprise storage cost can range from $5,000 - $10,000 (USD) per usable terabyte. Slower performing tier 3 storage ranges from $1,000 - $3,000 (USD) per usable terabyte.)

File and filegroup backups are similar in concept. Making a filegroup backup will back up each file in the file group, just as when you specify each file individually. When you create a file or filegroup backup, you can back up the entire file or filegroup or perform differential backups.

After a major disaster, utility companies have a great deal of work to do. In such cases, massive power outages can occur across a very large region. The utility companies have to mobilize and begin work on restoring power. They combine their efforts into restoring the critical areas first to help the region.

SQL Server offers similar choices for restoring data. An example of being able to quickly restore service to a VLDB that has multiple filegroups is performing piecemeal restores. Online piecemeal restores became available beginning with SQL Server 2005 Enterprise Edition and Developer Edition. A piecemeal restore lets you make the database partially available by restoring the PRIMARY file group first. Secondary filegroups can then be restored at a later time. With Enterprise Edition, you can restore the secondary file groups without taking the PRIMARY offline. With Standard Edition, you can still perform a piecemeal restore, however restoring the secondary filegroups is an offline operation.

You'll find several methods and justifications for using files or filegroups with SQL Server. The intent of this chapter is not to give a lesson in partitioning the database, but to present the methods to back up and restore using file and filegroup backups.

> **NOTE:** For the exercises in this chapter, you need to create a folder called Backups on the local C: drive (C:\Backups). You will also need a recent copy of the sample AdventureWorks database, which you can download at http://msftdbprodsamples.codeplex.com/releases/view/93587. This book uses AdventureWorks 2012, which is renamed to AdventureWorks. You will need to have run the SetupScript02.sql file, which will create a new database, LinchpinPress. You can find all scripts mentioned in this chapter in the Book Series section at www.LinchpinPress.com.

You will need to run a stored procedure that will create yearly filegroups for the years 2008 through 2014. Next, the stored procedure will create new data files per year and will assign each file to its respective filegroup. Last, it will use a partition function and move the data in the SalesInvoice table within the LinchpinPress database, based on the OrderDate column. Since the SalesInvoice data will not update previous years, set the filegroups <= 2014 to read only.

A database can have many filegroups, but let's start by showing that most databases start with a single filegroup called PRIMARY and a single MDF file and LDF file.

If you haven't run SetupScript02.sql, do so now. You will need to right click on Databases and chose refresh for the database LinchpinPress to appear. Next, right click on **LinchpinPress** and select **Properties** (as you see in Figure 7.1).

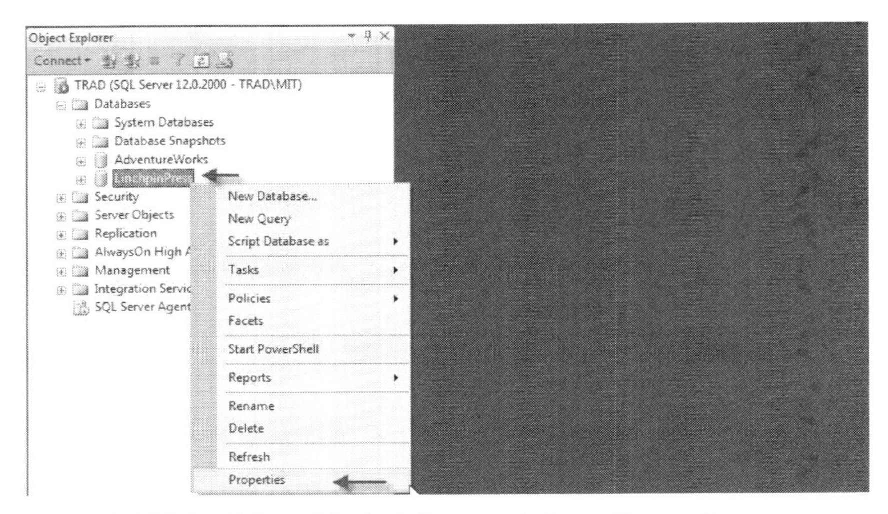

Figure 7.1 Right click on LinchpinPress and choose Properties

Next, click on the **Files** page, which you see in Figure 7.2. You'll see that you have only two files: your LinchpinPress data file in the PRIMARY filegroup and the LinchpinPress_log file.

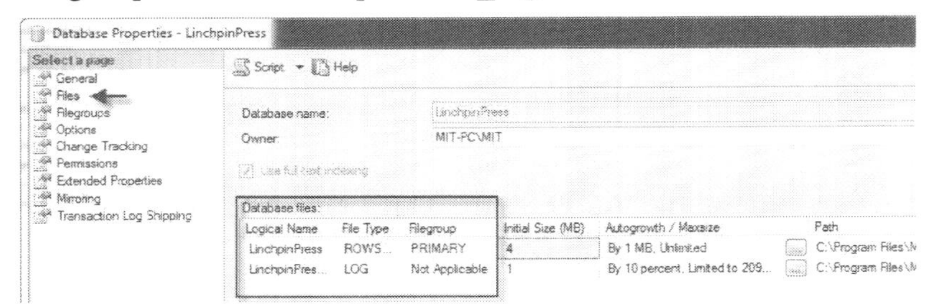

Figure 7.2 The Files page shows two database files, Primary data and the LinchpinPress log file

If you click on the **Filegroups** page (as in Figure 7.3), you'll see that you only have one filegroup called **PRIMARY**.

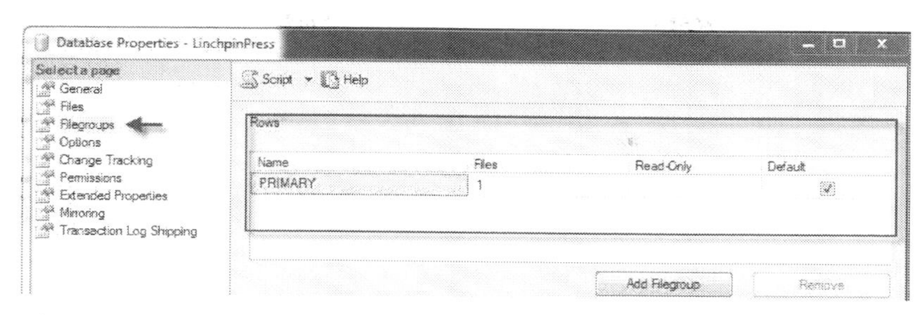

Figure 7.3 Filegroups page: The default LinchpinPress database has one file group

Click **OK** or **Cancel** to get back to the main SSMS page. Now that you know how a default database is configured, run the stored procedure that will create the additional filegroups and data files and also partition your data:

```
EXEC LinchpinPress.[dbo].[Chapter7]
```

Now repeat the steps above to review the files and filegroups, and you'll see the additional files and filegroups (as in Figure 7.4 and Figure 7.5).

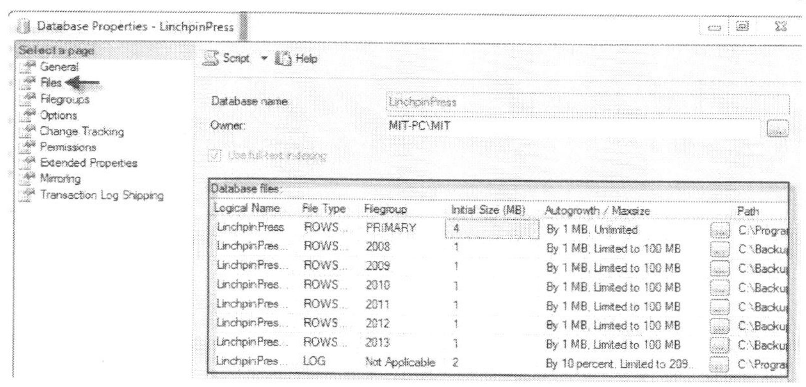

Figure 7.4 Additional files after partitioning

Click on the **Filegroups** page (as in Figure 7.5).

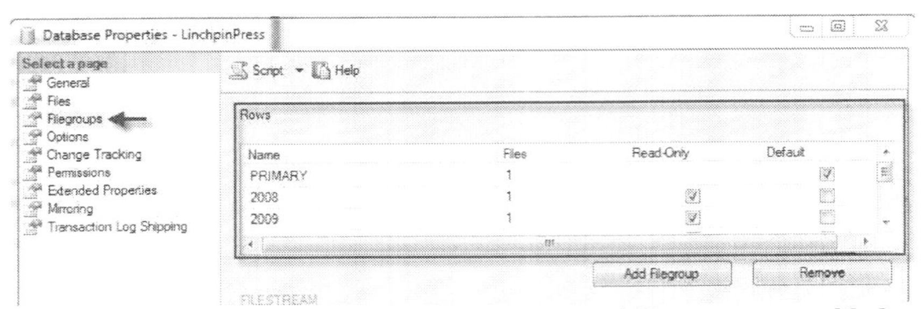

Figure 7.5 Filegroups page showing the number of filegroups you added

You are now ready to start performing backups and restores of your files and filegroups.

Using the Graphical UI

Creating a full backup of a database with multiple filegroups is similar to what you learned in Chapter 2. First, right click on the database, and choose **Tasks** > then the **Back Up…** option. Do this now with LinchpinPress (as shown in Figure 7.6).

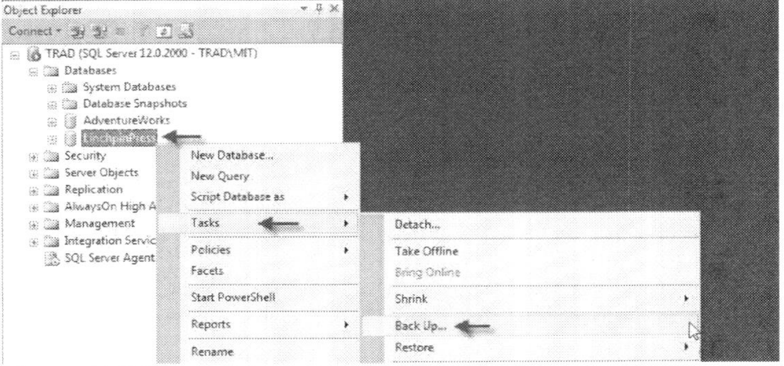

Figure 7.6 Right click on LinchpinPress > Tasks > Backup

Once you have chosen **Back Up…** , the **General** dialog box appears and you can choose among a number of selections. For now, do not make any changes until instructed. If you select **OK** at this point, you make a full backup of the database into the default backup location for the SQL Server instance.

In Figure 7.7, the first option is the database you want to back up. Next is the backup type. In this case, select FULL (as in Chapter 2). Alternatively, you can make a differential, which was covered in Chapter 3. You also see an option for **Copy-only Backup**, which was covered in Chapter 6.

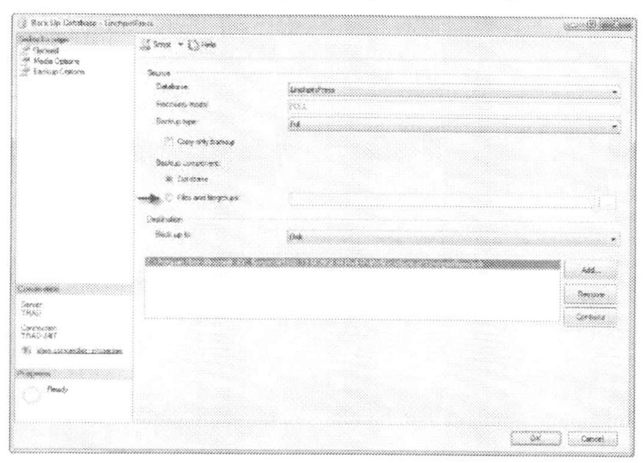

Figure 7.7 Check the radio button for Files and filegroups

For **Backup component**, select **Files and filegroups**, which will display a screen for you to select which file or filegroup to back up (as shown in Figure 7.8).

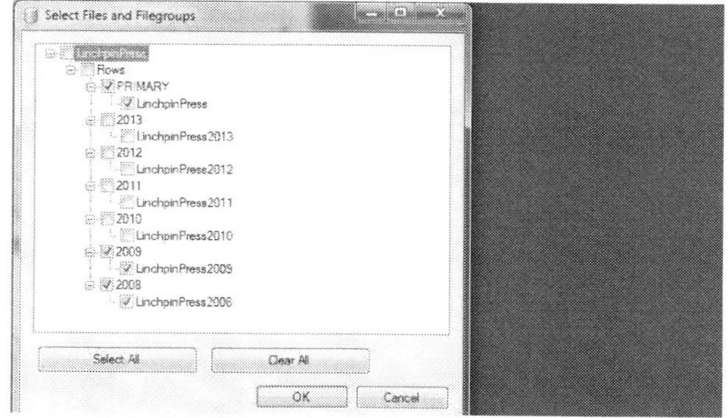

Figure 7.8 Select the files or filegroups you want to back up

The next item is the destination and name of the backup file. The default path is set at the instance level. For SQL Server 2014, the default location is:

C:\Program Files\Microsoft SQL Server\MSSQL12.MSSQLSERVER\MSSQL\Backup\.

While doing these steps, click the **Files and filegroups** radio button of the **Backup component** section. Choose to back up the LinchpinPress, LinchPress2008, and Linchpin2009 files. Choose the PRIMARY data file because you must restore the primary to bring the database online.

Click **OK** to complete the selection of files. You will now be back to the screen from Figure 7.7.

To change the backup path, click **Remove** > **Add** and type the path and name to use for the backup file (as you see in Figure 7.9).

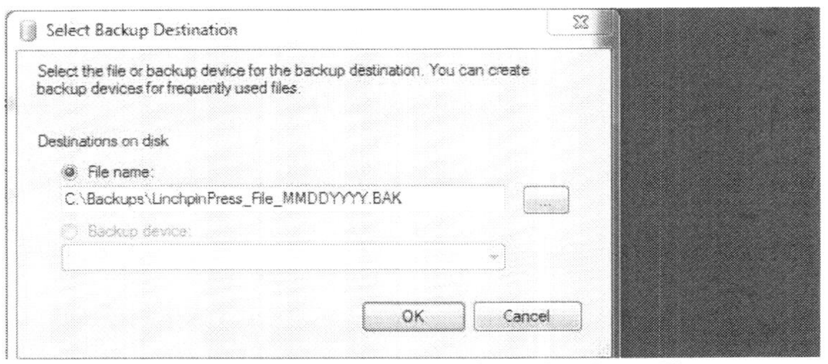

Figure 7.9 Type the path and name of the database file

As with all other backups, a best practice is to include a date and time, in addition to the backup type, in the backup name. Because there are many types of backups, it's helpful to have the name of the database in the backup file name. In this example, change the backup file name to:

C:\Backups\LinchpinPress_File_MMDDYYYY.BAK.

In the real world, if you had jobs that backed up different files on different days, it would be beneficial for the name to include information about which files were being backed up, as well.

Click the **Media Options** page, and you will see that you have several more options than you did in Chapter 2 for full backups. In Figure 7.10, you see the default values. **Overwrite media** defaults to **Append to the existing backup set**.

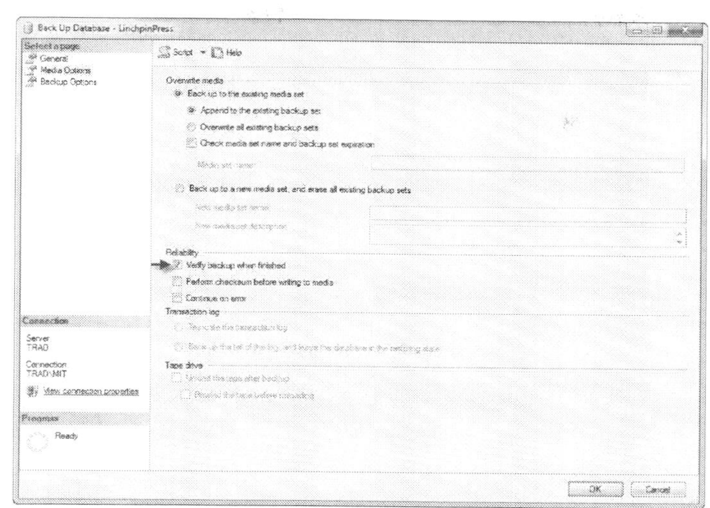

Figure 7.10 Additional options for backups

As with full backups, you need to verify file and filegroup backups by restoring them regularly. I recommend you select **Verify backup when finished**. However that does not mean the backup file is 100 percent valid. If the backup solution includes file and filegroup backups, the restore validation should also include restoring from them.

Click Backup Options (as shown in Figure 7.11), and you'll see that compression and encryption is also an option with **File and filegroup backups** and it performs the same way as it does with full backups.

With filegroup backups, it is especially important that everyone in the database group be familiar with how to restore. It would also be a great idea to have a documented process for restoring these environments.

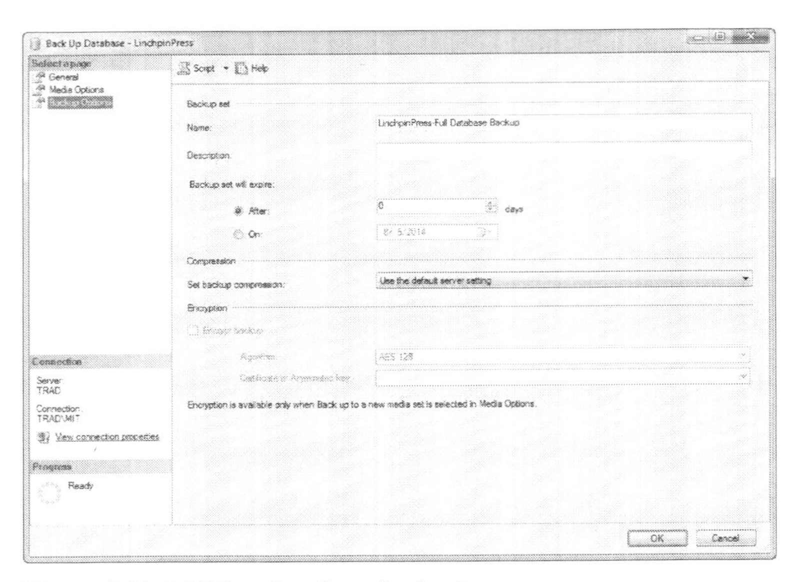

Figure 7.11 Additional options for backups

Once you've made the selections, click **OK** to perform the backup to: C:\Backups\LinchpinPress_File_MMDDYYYY.BAK.

Using T-SQL

The preferred method for making backups is to use T-SQL. Nowhere is the preference for T-SQL scripts more important than with file and filegroup backups. As you noticed in the previous section, we had to check several boxes to successfully make a backup. Having all these settings in a script makes it much easier for you to recreate the same backup consistently.

The syntax is straightforward.

```
BACKUP DATABASE” DB_NAME <file or filegroup> TO backup_device
WITH <OPTIONS>
```

In this example, include _TSQL at the end of the file name so that you do not append to the file you created using the graphical UI. Execute the following code now:

```
BACKUP DATABASE [LinchpinPress]
  FILEGROUP = 'PRIMARY',
  FILEGROUP = '2008',
  FILEGROUP = '2009'
  TO DISK =
'C:\Backups\AdventureWorks_File_MMDDYYYY_TSQL.BAK'
```

NOTE: This will only work if you ran the EXEC LinchpinPress.[dbo].[Chapter7] code from earlier.

Using the Graphical UI to Restore

Using the graphical UI is not an option for a piecemeal restore. It can only be done using code.

Using T-SQL to Restore

T-SQL is the only option for restoring one file or filegroup at a time. You can use the graphical UI to perform a restore of all filegroups, which is essentially a full restore.

To bring the database online, you must have a transaction log backup, unless you perform a partial restore. For this example, use the NO_RECOVERY option, which will put the database in a restoring state. This readies the database so you can begin the restore.

To also bring the entire database online, perform a full backup of the LinchpinPress database. The sequence of events will be:

1. Back up the entire database
2. Back up your primary, 2008 and 2009 filegroups
3. Insert a new record

4. Make a transaction log backup with norecovery to set the database in a recovering state
5. Restore your PRIMARY, 2008 and 2009 filegroups

Begin by executing the following statements:

```
USE MASTER
GO
BACKUP DATABASE LinchpinPress TO DISK =
  'C:\Backups\LinchpinPress_Full_FileDemo_MMDDYYYY.BAK'
GO
BACKUP DATABASE [LinchpinPress]
  FILEGROUP = '2008',
  FILEGROUP = 'PRIMARY',
  FILEGROUP = '2009'
TO DISK =
'C:\Backups\LinchpinPress_File_MMDDYYYY_TSQL.BAK'
WITH INIT
GO
INSERT LinchpinPress.[dbo].[SalesInvoice] ([InvoiceID],
[OrderDate], [PaidDate], [CustomerID], [Comment])
VALUES (9999, getdate()-1, getdate(), 777, NULL)
GO
BACKUP LOG LinchpinPress TO DISK =
  'C:\Backups\LinchpinPress_File_TLOG_TAIL.TRN'
WITH NORECOVERY
GO
RESTORE DATABASE LinchpinPress FILE = 'LinchpinPress'
FROM DISK =
'C:\Backups\LinchpinPress_File_MMDDYYYY_TSQL.BAK'
WITH PARTIAL, REPLACE, NORECOVERY
GO
RESTORE LOG LinchpinPress FROM DISK =
  'C:\Backups\LinchpinPress_File_TLOG_TAIL.TRN'
WITH RECOVERY
GO
RESTORE DATABASE LinchpinPress FILE = N'LinchpinPress2008'
FROM DISK =
  N'C:\Backups\LinchpinPress_File_MMDDYYYY_TSQL.BAK'
WITH RECOVERY
GO
RESTORE DATABASE LinchpinPress FILE = N'LinchpinPress2009'
```

```
FROM DISK =
  N'C:\Backups\LinchpinPress_File_MMDDYYYY_TSQL.BAK'
WITH RECOVERY
GO
```

> **NOTE:** This code will only run if you have done all the backup steps and remembered to change the file names of MMDDYYYY to what you used.

You can now check the state of each of your database files by executing the following script:

```
SELECT name, state_desc
FROM LinchpinPress.sys.database_files
```

In Figure 7.11, you can see that the LinchpinPress, LinchpinPress2008, LinchpinPress2009 files and the log are ONLINE, while the rest are RECOVERY_PENDING.

Results | Messages

	name	state_desc
1	LinchpinPress	ONLINE
2	LinchpinPress_log	ONLINE
3	LinchpinPress2008	ONLINE
4	LinchpinPress2009	ONLINE
5	LinchpinPress2010	RECOVERY_PENDING
6	LinchpinPress2011	RECOVERY_PENDING
7	LinchpinPress2012	RECOVERY_PENDING
8	LinchpinPress2013	RECOVERY_PENDING

Figure 7.12 The LinchpinPress, LinchpinPress log, LinchpinPress2008, and LinchpinPress2009 files are ONLINE

In the real world, this is where things would get interesting if the system is not set up to support a partial restore. For instance, suppose the users have an interface where they can select a date range to search on, and they put in a date such as >= '2008-01-01'. What happens when the query is executed? Go ahead and run the following query:

```
SELECT * FROM LinchpinPress.dbo.SalesInvoice
WHERE OrderDate >= '2008-01-01'
```

This gets an error message similar to the one below, which states the filegroup 2010 cannot be accessed because it is offline. This is because you restored the 2008 and 2009 files, which can be read, but 2010 is offline.

```
Messages
Msg 679, Level 16, State 1, Line 1
One of the partitions of index '' for table

                                        0 rows
```

In such situations, it's helpful to limit the users' ability to search outside the boundary of data that's online. In some cases, an application may be using a VIEW you can alter to limit the date range. Other systems might use a stored procedure. In most cases, you may just have to tell users to stay within a date boundary.

If you're using a date boundary, how does the same result set look? Execute a few more queries to find out:

```
SELECT *
FROM LinchpinPress.dbo.SalesInvoice
WHERE OrderDate >=
  '2008-01-01' AND OrderDate <= '2009-12-31'

SELECT *
FROM LinchpinPress.dbo.SalesInvoice
WHERE OrderDate >= '2014-01-01'
```

Each query returns a set of results with no errors. Since you're restricting date ranges to data that resides in your ONLINE filegroups, SQL Server can find your data and return it without error.

Let's restore the remaining filegroups to bring the database 100 percent back online. Since SQL Server Developer Edition has the same features as Enterprise Edition, this is an ONLINE operation. Execute the following statements:

```
RESTORE DATABASE LinchpinPress
  FILE = N'LinchpinPress2010',
  FILE = N'LinchpinPress2011',
  FILE = N'LinchpinPress2012',
  FILE = N'LinchpinPress2013'
FROM DISK =
  'C:\Backups\LinchpinPress_Full_FileDemo_MMDDYYYY.BAK'
WITH RECOVERY
```

If you now run the script to view the **state_desc**, you'll see all the filegroups ONLINE (as in Figure 7.13).

	name	state_desc
1	LinchpinPress	ONLINE
2	LinchpinPress_log	ONLINE
3	LinchpinPress2008	ONLINE
4	LinchpinPress2009	ONLINE
5	LinchpinPress2010	ONLINE
6	LinchpinPress2011	ONLINE
7	LinchpinPress2012	ONLINE
8	LinchpinPress2013	ONLINE

Figure 7.13 All filegroups are now ONLINE

You can now query the database again for data >= 2008, as you did before. This time you will not receive an error. Run the following script to verify:

```
SELECT * FROM LinchpinPress.dbo.SalesInvoice
WHERE OrderDate >= '2008-01-01'
```

Summary

Being able to restore portions of a database can have a huge advantage in a disaster situation. Think about very large companies that process tens of thousands of monetary transactions per minute. If they generate $100,000 an hour in revenue from their online system, then they are losing that much money every hour the system is down. Every hour they're down is costing more than someone's entire year's salary. Being able to get the primary data online to service the customers is critical.

Piecemeal and partial restores give you that ability, but such restores come with a price of being much more complex to manage. If the organization utilizes file or filegroup backups, then you need to dedicate time to practicing the restore of that system until you are extremely comfortable.

File or filegroup backups enable you to perform backups of VLDBs more quickly. This approach can ensure VLDB backups occur during maintenance windows so the I/O associated with backups does not impact users during production hours. In addition to providing flexibility with backups and restores, this approach enables you to place different files or filegroups on different sets of disks. Having many disks working for you helps improve disk I/O performance.

Before immediately jumping to the conclusion that you should partition or split a database into multiple file groups, check to make sure you are not storing data that has aged past the retention period for that type system. Many systems log activity that is no longer needed and can be routinely purged.

When performing a file or filegroup restore, you must have the accompanying transaction log backups in order to roll forward each file backup if you're using the full recovery model and if the file groups are read/write.

When you're making these decisions, you have to be careful if you have third-party products associated with your database. Have you ever seen a tag on a product that says, "Warranty void if seal is broken"? It's OK to use the product but certain areas of the product are not for the consumer to touch.

As database administrators, you support databases for many third-party products. The vendors that provide these products create their databases to support their products. You are tasked with making sure the data is backed up, indexes are optimized, and anything else you can do to make the system perform as well as it can.

In some cases, the license agreement specifies that only your vendor is authorized to change certain areas. Most vendors will allow DBAs to create additional indexes to help the system run better. However, this allowance comes with the understanding that those indexes may be lost when the vendor upgrades the application in the future. Some vendors drop indexes and recreate them during their upgrade process. When you start looking at partitioning data and splitting data across multiple filegroups, these vendors may not understand or support your decision. They may no longer support the system if you partition it. This can put you in a very bad situation.

A good practice is to work with the vendors on anything you do that will alter their database. In countless cases, DBAs have presented vendors with solutions to tune or tweak their systems, and the vendors have adapted these solutions to work with their products. At the end of the day, the vendor is just another group of technologists trying to earn a living, like any of us.

Points to Ponder
File and Filegroup Backup

1. Piecemeal restores allow you to bring up the most critical parts of your database first rather than waiting for the entire database to be fully restored.
2. To do a piecemeal restore, SQL Server does file and filegroup restores.
3. File and filegroup restores can only be applied to the databases in which they belong.
4. You cannot attach a single file of a database if the database has multiple data files. If SQL Server cannot locate every file of the database, the attach process fails.

Review Quiz – Chapter Seven

1. Online piecemeal restores are available in which editions of SQL Server?

a. Standard Edition
b. Express Edition
c. Enterprise Edition
d. Developer Edition

2. VLDB is the acronym for which phrase?

a. Very Large Database
b. Voluminous Long Data Block
c. Variable Elongated Database
d. Very Large Dense Backup

Answer Key

1.) Online operations such as a piecemeal restore are an Enterprise Edition feature and are also available in Developer Edition; therefore (c) and (d) are both correct answers.

2.) VLDB is the acronym for Very Large Database; therefore (a) is the correct answer.

Chapter 8. Backing Up System Databases

People take many things in life for granted, such as running water, electricity, and food being available in supermarkets. But because you can't always depend on the things you take for granted, you have to plan for the worst case: When bad weather is coming, people stock up on water and other supplies; some people purchase generators for emergencies; most people keep some supply of extra food in their pantry and refrigerator.

Just as with the things you take for granted in your personal life, you can't always depend on systems such as SQL Server to operate normally at all times. You have to plan for the worst in your work life, as well as your private life.

So far, the emphasis in this book has been on backing up and restoring user (i.e., non-system) databases. Anyone who has spent time with SQL Server knows that the SQL Server installation includes several key system databases that have different uses. Figure 8.1 shows a list of the SQL Server 2014 system databases.

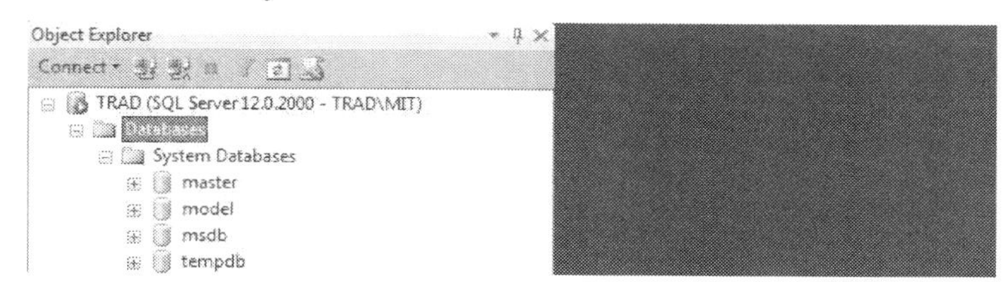

Figure 8.1 System Databases

SQL Server system databases perform all sorts of critical things for your environment that you may take for granted. These databases need special consideration when you're planning backup and recovery processes.

Each system database has a key function. The master database holds SQL Server's metadata and security objects such as users, DMOs, and more. Msdb holds all backup history, job history, SSIS packages, and more. Tempdb is used for many system functions such as snapshots, database mail, service broker, and all sorts of temporary objects. Model is the template database that all databases are based on when you create them. This is very important for tempdb since it is recreated based on the model database each time SQL Service is restarted.

Backing up system databases is not any different than backing up user databases, and you should include system databases in the regular backup routine. The only exception is tempdb. You do not need to back up tempdb—in fact, you cannot back up tempdb.

This leaves master, model, and msdb to worry about when you back up and restore. Since backing up these system databases is no different than backing up user databases, let's jump right into how to restore each database, beginning with master.

Restoring the Master Database

Most everyone today has a cell phone that has countless contacts stored on it. No one memorizes all those numbers, and most people don't keep a written version of their contact list and numbers. If you happen to lose your cellphone or it simply dies, how easily could you return to normalcy after replacing your phone? Fortunately, getting a new phone does not mean you need to retype all your numbers because most people are now storing their contacts on the provider's network, outside the physical phone itself. When you get a new phone, all your contacts are restored to the new phone.

The master database is sort of like your contact list for your phone in that it stores a list of users that have access to your databases. However, if you have to rebuild a SQL Server system and restore the user databases, you still have to do more than with your phone. Until you recreate the users on the server and grant permissions, your users cannot access the data.

Nobody should know all the users and passwords if you need to add them all to another server in the event of a disaster. To eliminate the risk of a person knowing all the users and passwords, you back up the master database as part of your normal backup routine.

The difficulty comes in with being able to restore the master database. To get started, you will need to back up the master database. So go ahead and execute the follow code:

```
BACKUP DATABASE MASTER TO DISK = 'C:\Backups\MASTER.BAK'
```

One of the most important considerations when restoring the master database is that the server you are restoring to must be running the same version of SQL Server as the original, down to the build number (patch level). If you are not in the habit of documenting the patch level of the SQL Server instances after every patch cycle, then don't worry.

It's easy to determine the build number by restoring the header of the master database. Execute the following script to find the Major.Minor.Build version:

```
RESTORE HEADERONLY FROM DISK = 'C:\Backups\MASTER.BAK'
```

The columns that reveal the version of SQL Server are SoftwareVersionMajor, SoftwareVersionMinor, and SoftwareVersionBuild. The columns are self-explanatory. SQL Server versioning is Major.Minor.Build. At the time this book was written, the version in the demos is 12.0.2000, which is SQL Server 2014 RTM.

Since SQL Server uses the master database, you can't just open up SSMS and type a RESTORE command and overwrite the system database. You must take specific steps. First, start the instance of SQL Server in single user mode to restore the master database. You can start SQL Server in single user mode by using either the startup parameter "-m" or using SQLCMD with "/m". Using SQLCMD is straightforward, and it is easy to follow along with. So for this demo, use SQLCMD.

To start SQL Server in single user mode using SQLCMD, you must first stop the SQL Server service of the instance you need to restore. Since you will be using SQLCMD, go ahead and open a command prompt and navigate to the BINN directory of the SQL Server instance.

To open a command prompt, click **Start** > **Run** >. Type **CMD**, and click **OK**. A black dialog box will open similar to what you see in Figure 8.2.

On some workstations or servers, you may need to open the command prompt in administrator mode.

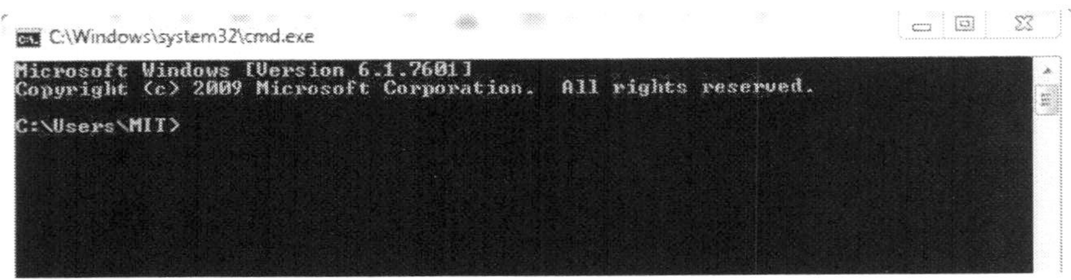

Figure 8.2 Illustration of a command prompt

Once you have the command prompt open, type **CD** and press **Enter**. This will take you to the root of the CD drive. (It's the equivalent of clicking the folder up button in Windows Explorer.)

Next you need to change directories to the BINN directory of SQL Server. On SQL Server 2014, when you're installing into the default directory, the BINN directory will be C:\Program Files\Microsoft SQL Server\MSSQL12.MSSQLSERVER\MSSQL\Binn. Other versions of SQL Server will have a similar path with a different version number. To change to this directory in the command prompt, you will need to type

CD Program Files\Microsoft SQL Server\MSSQL2.MSSQLSERVER\MSSQL\Binn

Press **Enter**.

Once you perform that step, you'll see a command prompt that looks similar to Figure 8.3.

Figure 8.3 Command Prompt in the BINN directory of SQL Server

To stop the default SQL Server instance, type **net stop mssqlserver**, and press **Enter**:

```
net stop mssqlserver
```

To start SQL Server in single user mode, type **net start mssqlserver /m**, and press **Enter**:

```
net start mssqlserver /m
```

Next type SQLCMD, and press **Enter**.

Now that SQL Server is running in single user mode, you can restore the master database by executing the following code. Type this all out on one line:

```
RESTORE DATABASE MASTER FROM DISK =
  'C:\Backups\MASTER.BAK' WITH REPLACE
```

Once you press **Enter**, you will be prompted with a second line.

Type **GO**, and press **Enter** (as you see in Figure 8.4).

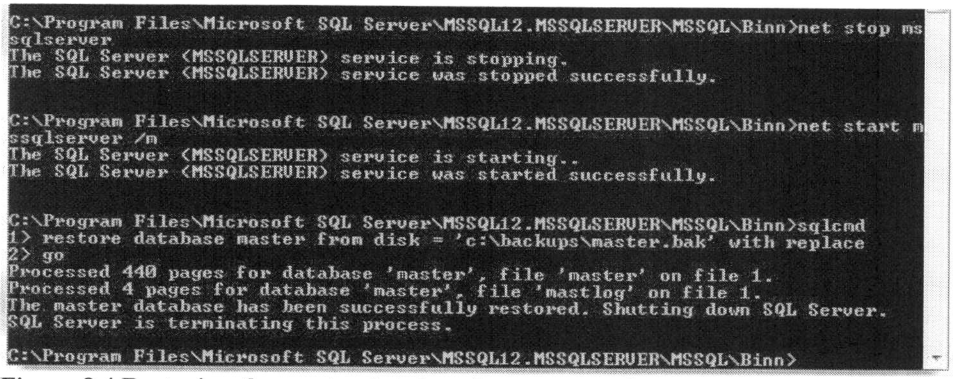

```
C:\Program Files\Microsoft SQL Server\MSSQL12.MSSQLSERVER\MSSQL\Binn>net stop ms
sqlserver
The SQL Server (MSSQLSERVER) service is stopping.
The SQL Server (MSSQLSERVER) service was stopped successfully.

C:\Program Files\Microsoft SQL Server\MSSQL12.MSSQLSERVER\MSSQL\Binn>net start m
ssqlserver /m
The SQL Server (MSSQLSERVER) service is starting..
The SQL Server (MSSQLSERVER) service was started successfully.

C:\Program Files\Microsoft SQL Server\MSSQL12.MSSQLSERVER\MSSQL\Binn>sqlcmd
1> restore database master from disk = 'c:\backups\master.bak' with replace
2> go
Processed 440 pages for database 'master', file 'master' on file 1.
Processed 4 pages for database 'master', file 'mastlog' on file 1.
The master database has been successfully restored. Shutting down SQL Server.
SQL Server is terminating this process.

C:\Program Files\Microsoft SQL Server\MSSQL12.MSSQLSERVER\MSSQL\Binn>
```

Figure 8.4 Restoring the master database in a command prompt

The master database will be restored, and the SQL Server instance will be shut down. To start SQL Service back up, type **net start mssqlserver** again, and press **Enter**.

Restoring the msdb Database

Big executives in large corporations have executive assistants to help them stay organized. The assistants will remind the executives of where they are supposed to be, important meeting times, a spouse's anniversary, birthdays, and anything else important. These executives become dependent on their assistants for many day-to-day tasks.

For your database servers, you get the same help. This helpful assistant is SQL Server Agent. SQL Server Agent is the scheduler for all the jobs that run on your SQL Server instance. These jobs may perform your database maintenance, run complex SSIS packages or other important processes.

Many of these jobs and the times they run are stored within msdb along with your backup history. Back up msdb by executing the following code:

```
BACKUP DATABASE MSDB TO DISK = 'C:\Backups\MSDB.BAK'
```

Most SQL Server automation will use msdb. That is generally true, unless you have every job scripted out and your SSIS packages are stored in a version control system somewhere. Aside from those exceptions, you will need to recover automation data from msdb. If you are doing an in-place restore or rebuilding the instance on another piece of hardware, you can just restore the msdb database to get it back online.

To restore msdb, you can follow either of the restore options in Chapter 2. This example will use T-SQL. Go ahead and execute the following script:

```
RESTORE DATABASE MSDB FROM DISK =
  'C:\Backups\MSDB.BAK' WITH REPLACE
```

Restoring the model Database

In today's industrialized society, few things are handmade. If you want to repeat or mass produce something, you typically create a template or a mold. This lets you easily and consistently produce copies of the object or repeat the task.

To help you easily and consistently reproduce databases with the characteristics you need, SQL Server includes the model database. New databases should all be created to a certain standard, a default configuration. The template that SQL Server uses is the model database.

Any changes that are made to the model database will be reflected in any new databases that are created. If you change the default size of the data file and log file or the percentage of growth, then when you create a new database it will have those changed settings.

The model database is critical for SQL Server since tempdb is recreated each time SQL Server is restarted. If you do not have a model database, SQL Server cannot create tempdb. Without tempdb, SQL Server will not start.

Backing up the model database is just like backing up master and msdb. Go ahead and make a backup using the following script:

```
BACKUP DATABASE MODEL TO DISK = 'C:\Backups\MODEL.BAK'
```

You can restore the model database just as you did previously with msdb. You can also use the graphical UI or T-SQL. For this exercise, use T-SQL to execute the following script:

```
RESTORE DATABASE MODEL FROM DISK =
  'C:\Backups\MODEL.BAK' WITH REPLACE
```

In the event of a disaster when you are rebuilding a database, restoring the model database is almost optional, unless you have made changes to the default install of model. Realistically, if you are following best practices, then surely you will have made a few changes to the default size and auto growth settings. However those can easily be changed manually during the restore. Most organizations account for those changes in their server build procedures.

Restoring the tempdb Database

In the 5th grade, my class got extra points for doing helpful things for the teacher each day. The job I did the most to receive my accolades was erasing the whiteboard at the end of each class. The teacher worked all day at putting data up on the board relating to our lessons. So why was he happy when I cleaned it off each day? It was because the next day, for a new class lesson, he would have a new work area with new data. Also, sometimes the teacher would erase half the board in the middle of the day to make some corrections to the data he was presenting. The whiteboard was like a thinking area for data in the teacher's real-time workflow.

SQL Server can also put its ideas and temporary work into something like its own type of whiteboard area. This data work area for SQL Server is the tempdb. None of its contents are officially stored or have any long-term relevancy to the other databases.

Since tempdb is recreated with each SQL Server restart, you don't need to back up and restore tempdb. As a matter of fact, backing up tempdb is not even an option. Go ahead and try by executing the following script:

```
BACKUP DATABASE TempDB TO DISK =
  'C:\Backups\TempDB.BAK'
```

You'll get the following error:

```
Messages
Msg 3147, Level 16, State 3, Line 1
Backup and restore operations are not allowed on database TempDB.
Msg 3013, Level 16, State 1, Line 1
BACKUP DATABASE is terminating abnormally.
                                                          0 rows
```

Summary

Over the years, reading blog posts, answering questions on forums, and speaking at SQL Server events, I've heard a common story. It involves data professionals who suffer a production crash that involves bringing up their systems on a new server. Once the data professional installs SQL Server and restores the user databases, none of the applications or users can connect to the restored database. The DBA quickly realizes all the users were stored in the master database and has to rush to restore that database, as well.

What those DBAs lack is knowledge and experience in restoring the master database. When the DBA tries to restore the master database, she finds that the install of SQL Server is not at the same patch level as the backup. So now she has to download updates to get it to the same level.

Once she overcomes that issue and gets master database restored, users and applications are able to connect.

At this point, the DBA feels good that the system is up. She basks in the glory that she saved the day.

However, the next day she comes to work and finds a new problem has been reported: None of the ETL jobs ran last night. The DBA forgot that all the jobs and packages are stored in the msdb database. So now she has to restore msdb and manually run each job to get the system caught up. Once all the jobs process, the system returns to normal and business goes on as usual.

It's important to keep in mind that each system database performs a unique service to SQL Server. Had this DBA practiced doing a full system restore, she would have saved her company hours of downtime and lost productivity.

System databases are a critical part of SQL Server, and you need to back up all but tempdb in your regular backup routine. You back up Master, model, and msdb, just like a user database. You also restore model and msdb like user databases.

You must restore the master database in single user mode, and the SQL Server instance must be at the same build level as the backup. This can mean extra steps when you're trying to recover a system.

The tempdb database is recreated each time SQL Server service is started; therefore it cannot be backed up and restored.

Points to Ponder Backing Up System Databases

1. A major part of the restore strategy for a SQL Server instance is recovering data that is stored in the Master and msdb system databases. If you can successfully restore a user database, but no users are configured with permissions, you have not successfully restored the system.

2. If the master database is damaged so much that you can't start up SQL Server, you can attempt a rebuild.
3. You have to use the same version of SQL Server for both the backup and the restore. This means that to restore, you need to have all the service packs and installation media available on the network. In times of a disaster, the Internet may not be available for you to download updates from MSDN.

Review Quiz – Chapter Eight

1. You must include tempdb in your backup routine.
   a. TRUE
   b. FALSE
2. The model database is the template for new database creations.
   a. TRUE
   b. FALSE
3. SQL Server must be in single user mode in order to restore the master database.
   a. TRUE
   b. FALSE

Answer Key

1. The tempdb database can't be backed up; therefore the answer is (b) FALSE.
2. SQL Server uses the model database as the template for new database creations. The answer is (a) TRUE
3. To restore the master database, the SQL Server instance has to be at the same build level as the database and you must start SQL Server in single user mode. The answer is (a) TRUE.

Chapter 9. Additional Best Practices

Anyone who owns a car knows owning and operating a car involves more than just knowing how to drive. You have to perform many maintenance tasks that your car may not alert you to. Regular maintenance helps keep your vehicle running smoothly and extends its life. You can shorten your car's life span if you neglect to perform some maintenance tasks such as changing the oil, rotating the tires, replacing air filters, and checking tire pressure.

Being responsible for SQL Server and recovering databases is like maintaining a car. You need to stay on top of maintenance tasks, or best practices, if you want a server that runs smoothly and you want to prolong your career. The following sections explain several areas that every DBA should be aware of.

DBCC CHECKDB

Corruptions happen in daily life (especially in government and politics). With technology, application programs have errors, hard drives crash, network switches can fail, and a whole host of other problems can develop.

Your databases sit on very large disks that are connected to complex servers. Any number of bad things can happen while data is being written to the disks. Although SQL Server is extremely good at doing error handling, it can't save you from errors 100 percent of the time. When it can't, the database, or a portion of it, can become corrupt.

To scan and check for potential corruption, run DBCC CHECKDB on your databases. This command checks the integrity of the database, both logical and physical. CHECKDB does this by running three checks:

1. CHECKALLOC checks the consistency of disk space allocation
2. CHECKTABLE checks the pages and structures of the table or indexed view

3. CHECKCATALOG checks the catalog consistency

CHECKDB also validates Service Broker data, the content of indexed views, and link level consistency of VARBINARY(MAX) data in the file system, using FILESTREAM.

I highly recommend that you regularly run DBCC CHECKDB against all your production databases. A best practice is to have this scheduled to run as part of the regular maintenance and to output the results to a file.

If corruption is detected quickly enough, you have a much better chance of recovering it from backup or being able to repair the corruption.

Crash Recovery

When a database is restored, the first thing that happens is the file is created and initialized. This can be time consuming if the database is large and instant file initialization is not enabled. Once the file is initialized, the data is copied. The final steps are the redo and undo portion, which is called crash recovery.

Any transactions that were written to the log file but not to the data file prior to the crash must be rolled forward to put the database in a consistent state. If any transactions were in flight and not committed before the crash, they must be rolled back.

Instant File Initialization

When you visit a financial institution to cash a check or withdraw money from your account, the tellers always count the money twice. They typically count it first for themselves and then count the money to you as they're handing it to you. This is how they guarantee that you are getting what you expect and that they don't accidently give you too much money and short their drawer. This act of counting out the money twice gives you a sense that the teller is trustworthy even though it lengthens the time of your transaction.

Similarly, when you create a database in SQL Server, that file is created on the NTFS file system, and SQL Server has to zero out each page in the file. This can be a very time-consuming task.

In SQL Server 2005 and above on Windows Server 2003 and above, you can enable instant file initialization. You turn this on by going into **Security Settings** > **Local Policies** > **User Rights Assignment** > **Perform Volume Maintenance Tasks** > and adding your SQL Server service account. Now when you create a new database, expand a database file, or drop and restore a database. SQL Server no longer has to zero out each page of the data file.

Summary

Checking for corruption in your databases is very important. Corruption can occur for many reasons. But 95 percent of the time, it is related to a hardware issue. Remember, SQL Server is very forgiving and will back up a corrupt database. Just because your database can be backed up and users can query the database does not mean everything is OK. You can protect yourself by scheduling DBCC CHECKDB to run as an integral part of database maintenance.

Understanding crash recovery is important. You need to know what process a database goes through after being restored or recovered from a server crash.

Implementing instant file initialization can drastically improve your recovery time when you're restoring a database to a new server, or if you're restoring after a database has been dropped.

Points to Ponder: Additional Best Practices

1. DBCC CHECKDB is very well documented on MSDN. If any errors return after you run CHECKDB, you can find arguments (additional options) that you can use with CHECKDB to dig further into the problem or possibly correct the issue.
2. Dealing with database corruption is usually not a pleasant experience. Having proper backups is the key to being able to repair and recover from corruption. When corruption occurs, if you don't have the data backed up before the corruption occurred, then in most causes that data is lost. If you are not regularly checking for corruption issues, your backup retention may not be sufficient for you to go far enough back in time to recover the data.
3. Instant file initialization can drastically decrease the amount of time it takes for tempdb to be recreated when SQL Server starts.

Review Quiz – Chapter Nine

1. DBCC CHECKDB performs what function in SQL Server?
   a. It checks to make sure the database has been backed up
   b. It checks for consistency to find corruption
   c. It reorganizes indexes
   d. It checks to ensure database mirroring is in sync
2. When SQL Server makes backups, it will check for corruption and will *not* back up a corrupt database.
   a. TRUE
   b. FALSE

Answer Key

1. DBCC CHECKDB checks for corruption. Therefore answer (b) is correct.
2. SQL Server can back up corrupt databases, so answer (b) is correct.

Chapter 10. Encryption

Security is important. Our banks lock cash and coins away in a very secure safe. The United States government stores gold in one of the most secure places in the world. Your organization requires you to log on to the network and probably has some form of physical security to access the building.

Data security is equally important. One type of security is to require authentication for users to access your databases. Encryption is another important means of securing your data. To secure the data stored in SQL Server, on most higher-end enterprise Storage Area Networks (SANs), your data that is stored on the disk is encrypted.

Note, however, that with versions older than SQL Server 2014, native SQL Server backups are not encrypted unless you explicitly use Transparent Data Encryption (TDE). What does lack of encryption mean? Anyone with read access to the drive that is holding your backups can make a copy and restore that database to another SQL Server instance running the same version as or higher than the copied database. Once that person restores the database, if he or she is a Server Administrator (SA) on the new server, that person has access to all the data.

Who are the domain or local server administrators on your database servers? Any one of them can make a copy of that database.

What can you do to protect your backups? The easiest thing is to back up to an encrypted disk or disk share if you have that option available in your organization. If you are using TDE, then your backups of databases encrypted with TDE are also encrypted. If neither of these options is available and you are using a version of SQL Server earlier than 2014, several third-party tools are available. Many of these tools also support backup compression, as well as the option to encrypt the backup files.

In SQL Server 2014, backup encryption is included. Now when creating a backup, you can chose to encrypt the file. To encrypt during the backup, you will have to specify an encryption algorithm and an encryptor.

In this case, you don't have to worry about someone gaining access to your encrypted backup files (even if someone has access to the backup), unless that person also gets the encryption key or password. Without the key, no one can restore and access the data.

Prerequisites for Encryption

To start creating backups with encryption, you will need a database Master Key, as well as a certificate, or asymmetric key, which is available on the instance of SQL Server.

You can easily create a Master Key if your server does not already have one. Open a new query window and type the following command, specifying your own password.

```
-- Creates a database master key.
-- The key is encrypted using the password "LinchpinPress"
USE master;
GO
CREATE MASTER KEY ENCRYPTION BY PASSWORD =
'LinchpinPress';
GO
```

Next you will have to create a certificate, or asymmetric key, for the instance if one does not already exist. To create a certificate, use the following script. You can change the name of the certificate and subject to meet your needs.

```
--Creates a new certificate for your instance

Use Master
GO
CREATE CERTIFICATE LinchpinPressEncryptCert
   WITH SUBJECT = 'LinchpinPress Backup Encryption
Certificate';
GO
```

If you backed up the database now with encryption, you would get the following warning:

```
Warning: The certificate used for encrypting the database
encryption key has not been backed up. You should immediately
```

```
back up the certificate and the private key associated with the
certificate. If the certificate ever becomes unavailable or if
you must restore or attach the database on another server, you
must have backups of both the certificate and the private key or
you will not be able to open the database.
```

Backing up the Master Key and server certificate is very similar to backing up a database. To back up your Master Key, simply update the file name and password below and execute the following script:

```
BACKUP MASTER KEY TO FILE = 'C:\Backups\MasterKey'
    ENCRYPTION BY PASSWORD = 'C0MplexP@$$w0rd'
```

You can use the next script to back up your server certificate. You will need to update the certificate name with the one you chose, specify the backup file name of the certificate, specify the file name of the private key backup file, and chose a password to encrypt the backup keys with.

```
BACKUP CERTIFICATE LinchpinPressEncryptCert
      TO FILE = 'C:\Backups\LinchpinPressEncryptCert'
    WITH PRIVATE KEY ( FILE =
'c:\Backups\LinchpinPressEncryptCertkey' ,
    ENCRYPTION BY PASSWORD = 'C0MplexP@$$w0rd' );
```

It is very important to store these passwords and backup files someplace accessible and safe. You will need them if you have to restore the database

> **NOTE:** For the exercises in this chapter, you need to create a folder called Backups on the local C: drive (C:\Backups). You will also need a recent copy of the sample Adventureorks database, which you can download at http://msftdbprodsamples.codeplex.com/releases/view/93587. In this book, we are using AdventureWorks 2012, which we renamed to AdventureWorks. You will need to have run the SetupScript01.sql file. You can find all scripts mentioned in this chapter in the Book Series section at www.LinchpinPress.com.

to another server.

How to Back Up with Encryption

For this example, you will create a full backup, just like you did in Chapter 2.

To make a full backup using the SSMS graphical UI, you need to follow a few simple steps. First, right click on the database, and choose **Tasks** >. Select the **Back Up…** option. Do this now with AdventureWorks (as Figure 10.1 shows).

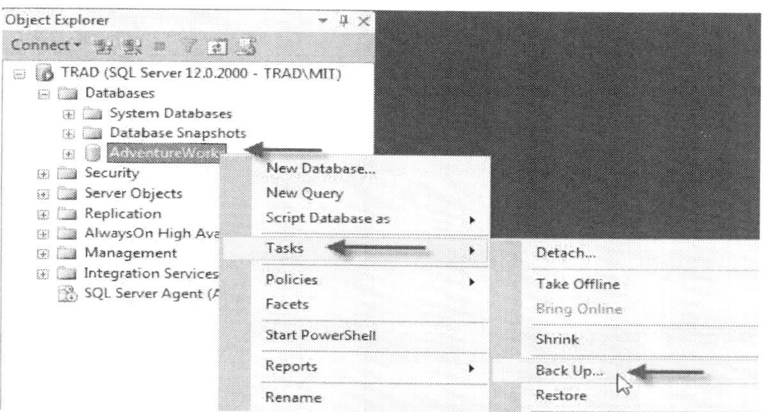

Figure 10.1 Right click on the database, choose Tasks, and then Back Up…

Once you have chosen **Back Up…** , the general dialog box appears where you can select from several options. Selecting **OK** at this point creates a full backup of the database into the default backup location for the SQL Server instance.

In Figure 10.2, the first option is to choose which database to back up. Next is the backup type you want to make. In this case, you want to choose **Full**. For **Backup component,** make sure **Database** is selected. The next item is the destination and name of the backup file.

Figure 10.2 General options for backup

The default path is set at the instance level. For SQL Server 2014, the default location is the following Path:

C:\Program Files\Microsoft SQL Server\MSSQL12.MSSQLSERVER\MSSQL\Backup\

In this example, you do not want to save in that location. To change this for the example, select the path to highlight it, and then click **Remove** >. Then click **Add** to make a new path.

Type C:\Backups\AdventureWorks_Full_Encrypt_MMDDYYYY.BAK, (as shown in Figure 10.3).

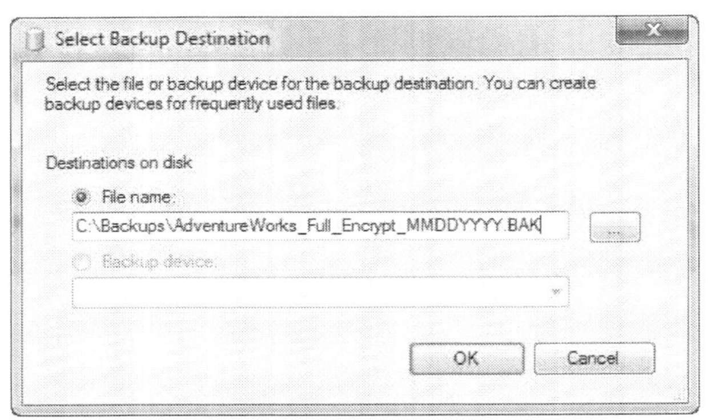

Figure 10.3 Type the path and name of the database file

A best practice is to include a date and time, as well as backup type, in the backup name. In this example, change the backup file name to AdventureWorks_Full_Encrypt_MMDDYYYY.BAK. Type the path and name of the database file. Replace MMDDYYYY with the actual date value in the form of Month Day Year. For example, if today is June 1, 2014, then the filename should be AdventureWorks_Full_06012014.BAK. This gives you a nice visual aid so you don't have to look at the timestamp on the file itself. Click **OK**. Note that when you're creating a custom backup job, you want to include HHMMSS in the name, as well.

The choices so far in the **General** page of the **Back Up Database** dialog are just a few of the options available. Just below **General**, click **Media Options** to see several more possible backup choices. Figure 10.4 shows the options you will need to change in order to back up with encryption.

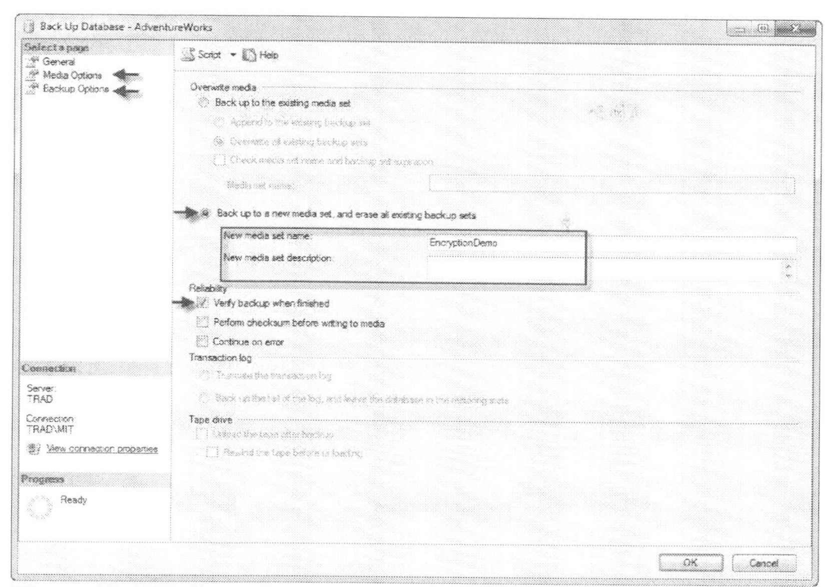

Figure 10.4 Media options for backups

Overwrite media defaults to **Append to the existing backup set**. To back up with encryption, you have to back up to a new media set. Since you have already taken multiple backups of AdventureWorks, you will have to specify a new media set name. Click **Back up to a new media set, and erase all existing backup sets**. In this exercise, you can chose EncryptionDemo as the name.

I recommend you always choose the option **Verify backup when finished.** However that does not mean the backup file is 100 percent valid. The only true way to fully verify a backup is to restore it.

Next, click on Backup Options. Figure 10.5 shows the options you will need to change in order to backup with encryption. Here you will check the box to **Encrypt backup**.

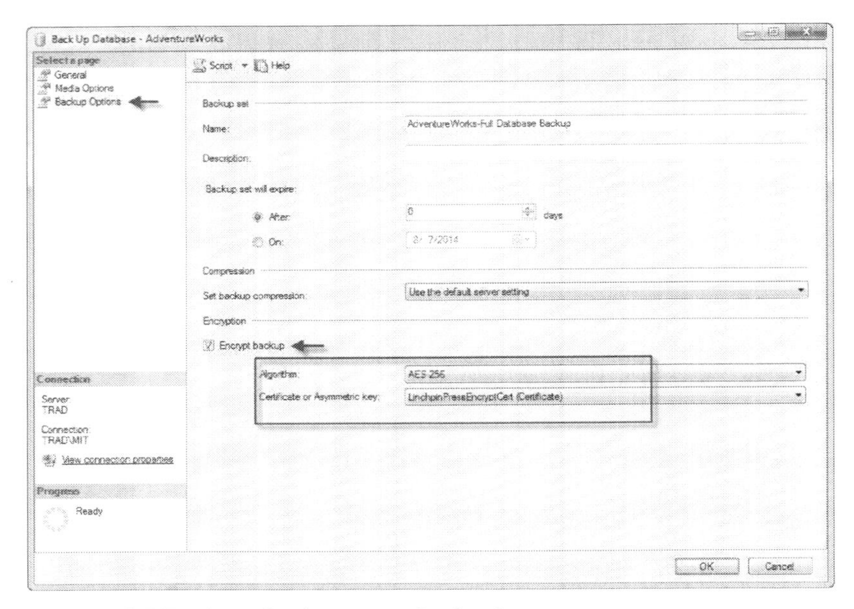

Figure 10.5 Backup Options page for backups

Compression

SQL Server 2008 Enterprise and above can also use compression. Using compression to back up means that backups will take up much less space and also take less time to complete because they require less device I/O.

If you are using a version of SQL Server that supports compression, you will see that option listed. For this exercise, use the default option from the following three options:

1. Use the default server setting
2. Compress backup
3. Do not compress backup

Encryption

Click the check box to Encrypt backup. The algorithm and certificate, or asymmetric key, options now enable up to choose you options. For the algorithm you have four choices

1. AES 128
2. AES 192
3. AES 256
4. Triple DES

The difference in the AES encryption algorithms is the size of the key lengths. If you are encrypting your database to meet a regulatory or compliance requirement, check the documentation regarding the key length requirement for your situation. The Triple Data Encryption Algorithm (TDEA or Triple DEA) has three key sizes: 168, 112, or 56 bits.

For this exercise, choose AES 256, and select your Certificate. Click **OK** to complete an encrypted full backup of the AdventureWorks database.

Using T-SQL to Encrypt

The syntax to back up with T-SQL code is straightforward:

```
BACKUP DATABASE DB_NAME TO <backup_device> WITH <options>
```

In this example, you will have some new options to specify. To use encryption, specify WITH ENCRYPTION and list the algorithm and server certificate to use.

Use the following script to back up the AdventureWorks database. You will need to update the server certificate to reflect the one you created earlier. As you have done in previous examples, replace MMDDYYYY with the actual date value.

```
BACKUP DATABASE AdventureWorks
TO DISK =
'C:\Backups\AdventureWorks_Full_Encrypt_MMDDYYYY_TSQL
.BAK'
WITH   COMPRESSION, ENCRYPTION
  (ALGORITHM = AES_256,
  SERVER CERTIFICATE = LinchpinPressEncryptCert)
```

Figure 10.6 shows the message confirming full backup and successful encryption.

```
Messages
Processed 24272 pages for database 'AdventureWorks', file 'AdventureWorks2012_Data' on file 1.
Processed 2 pages for database 'AdventureWorks', file 'AdventureWorks2012_Log' on file 1.
BACKUP DATABASE successfully processed 24274 pages in 5.296 seconds (35.808 MB/sec).
```

Figure 10.6 Full backup with successful encryption

Restoring an Encrypted Backup

To restore an encrypted database, you do not have to specify any encryption parameters. You do need to have the certificate or asymmetric key that you used to encrypt the backup file. This key or certificate must be available on the instance you are restoring to. Your user account will need to have VIEW DEFINITION permissions on the key or certificate.

If you are restoring a backup encrypted from TDE, the TDE certificate will have to be available on the instance you are restoring to, as well.

Summary

Encrypting backups provides great protection for preventing data breaches. If you are not encrypting your backups, discuss this with your team to ensure the risk is minimal and proper safeguards are in place.

Encrypting your backups adds additional risk for being able to restore. If you do not have your key or certificate you used to encrypt the backup, you will not be able to restore the backup. The same safeguard you're using to protect your data from being stolen could lock you out, as well. Make sure you have in place proper safeguards, such as storing your key in a backup location. Regularly test restoring your encrypted backups just like you should with any other backup.

Points to Ponder—Encryption

1. Protecting your backup files is just as important as protecting your databases. In some cases, it may be easier for a disgruntled employee to access your backup files than the database itself.
2. Protecting your certificate or asymmetric key is very important. Without the certificate or key you can't restore the backup.

Review Quiz – Chapter Ten

1. SQL Server 2014 provides two algorithms for encryption.
   a. TRUE
   b. FALSE
2. When restoring an encrypted backup, you do not have to specify any parameters in order to restore.
   a. TRUE
   b. FALSE

Answer Key

1. SQL Server 2014 has four algorithms for encryption. AES 128, AES, 192, AES 256, and Triple DES. Therefore answer (b) is correct.
2. You are not prompted to provide any parameters when restoring from an encrypted backup. Before you restore the database, the asymmetric key or certificate must be available on the SQL Server instance you are restoring to. Therefore answer (a) is correct.

Chapter 11. Back Up to Microsoft Azure Storage

Beginning with SQL Server 2012 SP1 CU2, you can back up your SQL Server databases to Microsoft Azure cloud storage. This is an excellent option to get your databases offsite. Many small and midsized businesses back up their data but store the backups in the same physical location as their servers. In the event of a fire, flood, tornado, or other natural disaster, such businesses could lose everything. It is important to have a secure offsite copy of your data.

Setting Up Your Azure Storage

To start backing up your databases to Azure Binary Large Object (BLOB) storage, you must first have an Azure account. You can visit http://azure.microsoft.com to set up your account.

Once you have an Azure account, you will have to create a storage account and then a container. To create your storage account, log in to the Management Portal, click on **STORAGE**, and then **CREATE A STORAGE ACCOUNT** (as you see in Figure 11.1).

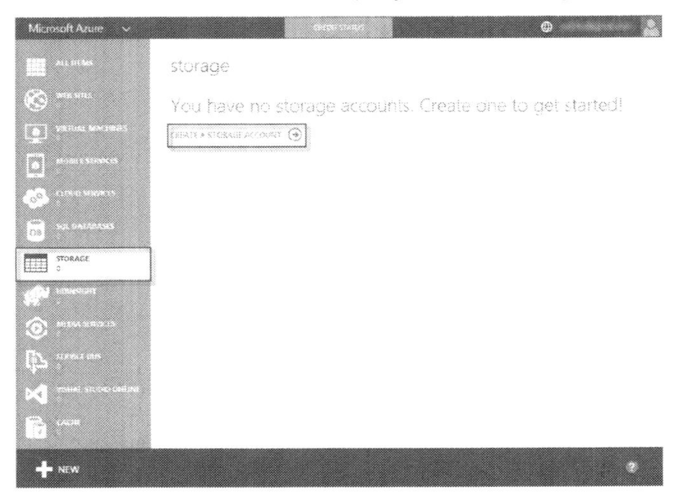

Figure 11.1 Microsoft Azure storage page

Click on **DATA SERVICES**, **STORAGE**, and then **QUICK CREATE**, as shown in Figure 11.2. You need to enter an address for your storage URL in all lowercase. Enter the name of your choice, and then click **CREATE STORAGE ACCOUNT**.

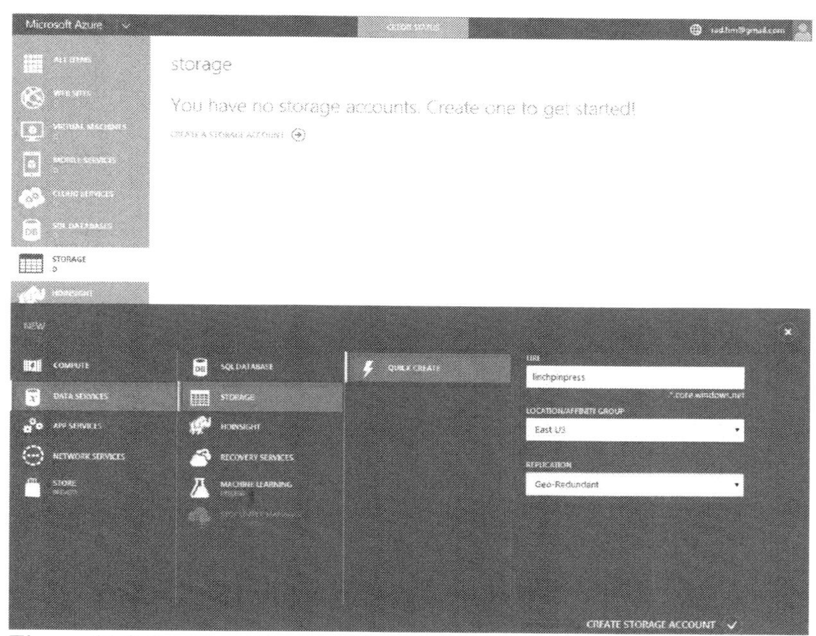

Figure 11.2 Create a Storage Account

Your account will be created and will display as Online once it is finished, as in Figure 11.3.

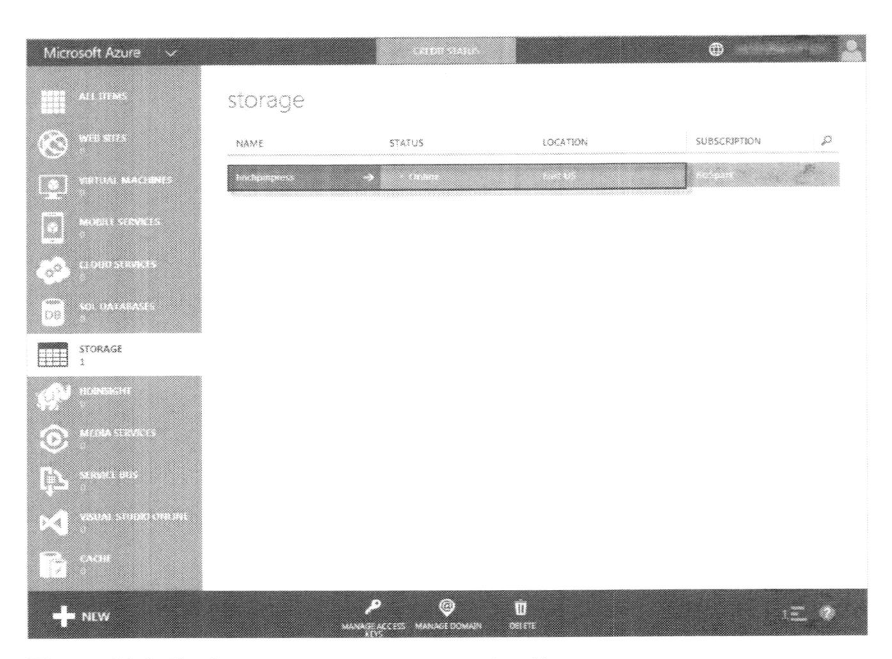

Figure 11.3 displays your storage account online

Within your storage account, you have to create a container to store the database backups into. Click on your storage name. This will take you to your storage dashboard. From here, you will need to click on **CONTAINERS**, as shown in Figure 11.4.

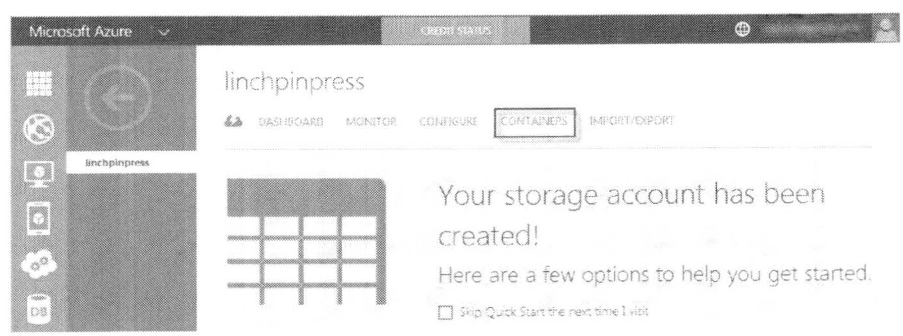

Figure 11.4 Storage dashboard

If this is your first time setting up a container, you will have to click on **CREATE A CONTAINER**. Do this now (as you see in Figure 11.5).

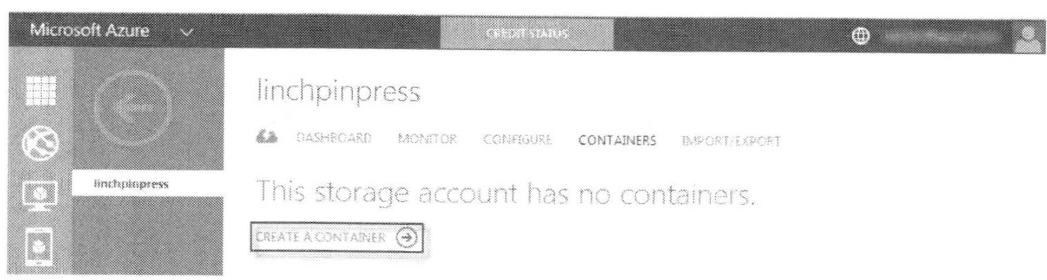

Figure 11.5 Click create a container

You will need to give this container a name, which must be all lowercase letters or numbers and can contain hyphens. Give your container a name, and click the checkmark to continue (as Figure 11.6 shows).

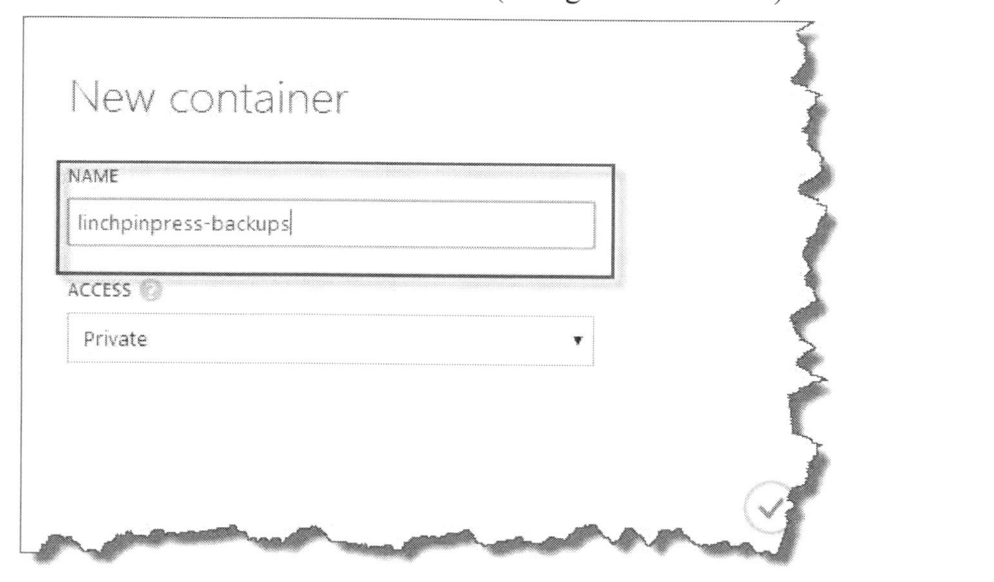

Figure 11.6 give your container a name

Once you create the container the screen will display your active containers (as in Figure 11.7).

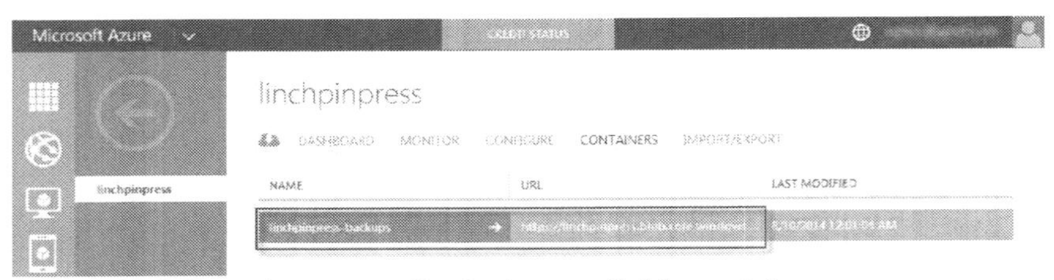

Figure 11.7 Container page displaying available containers

Next, click on **DASHBOARD** to take you back to your main storage page. You will see **MANAGE ACCESS KEYS** at the bottom center of the page (as in Figure 11.8). You will need your storage account name and primary access key to create the SQL Server credentials to start backing up to your Azure storage.

Figure 11.8 Manage Access Keys

Configuring your SQL Server Credential

Next, open SSMS, and expand **Security**. Right click on **Credentials**, and choose **New Credential** (as you see in Figure 11.9).

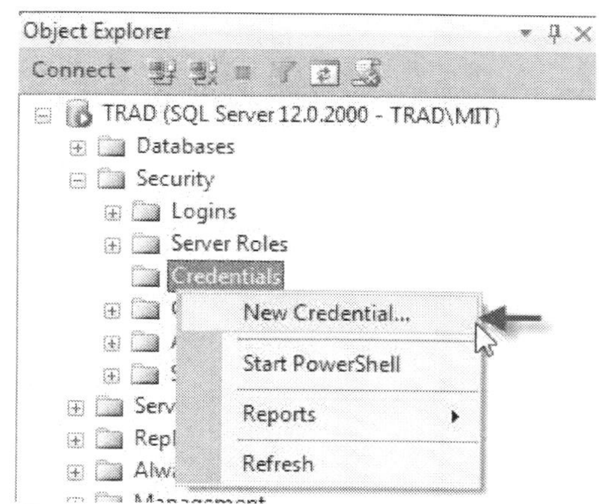

Figure 11.9 Right click on Credentials, and choose New Credential

Give your new credential a name. Identity will be your storage account name, and the password will be your primary access key. Enter your information and click OK (as in Figure 11.10).

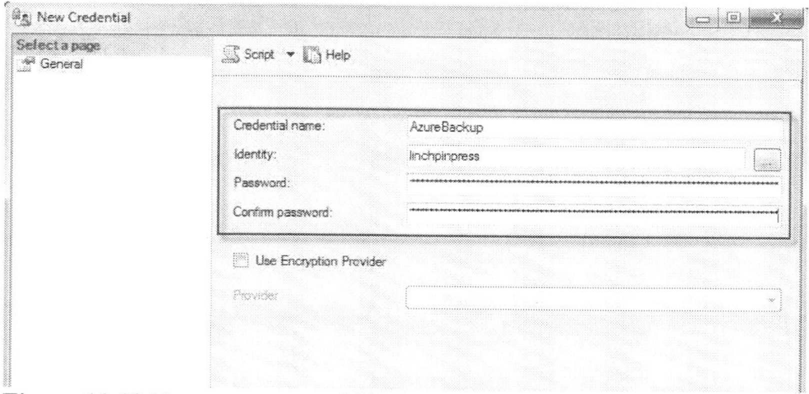

Figure 11.10 Name your new SQL Server credential

Backing Up to Azure Storage

To use the SSMS graphical UI to make a full backup, follow a few simple steps. First, right click on the database, and choose **Tasks >**. Select the **Back Up…** option. Do this now with AdventureWorks (as Figure 11.11 shows).

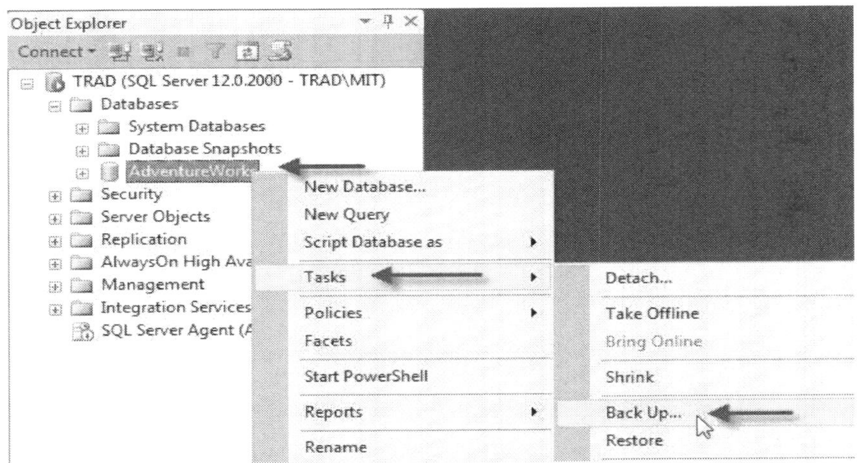

Figure 11.11 Right click on the database, choose Tasks, and then Back Up…

Once you have chosen **Back Up…** , the general dialog box appears where you can select from several options. Selecting **OK** at this point creates a full backup of the database into the default backup location for the SQL Server instance.

In Figure 11.12, the first option is to choose which database to back up. Next is the backup type you want to make. In this case, you want to choose **Full**. For **Backup component,** make sure **Database** is selected. The next item is the destination and name of the backup file. For **Back up to**: Choose URL. Provide the file name, select your SQL Server credential, and type the name you selected for your Azure storage container. The URL prefix will prefill for you.

If you want to make any other selections, such as compression, copy-only, etc., you can make those choices, as well.

Click **OK**, and you will begin your backup to Azure storage (as you see in Figure 11.12).

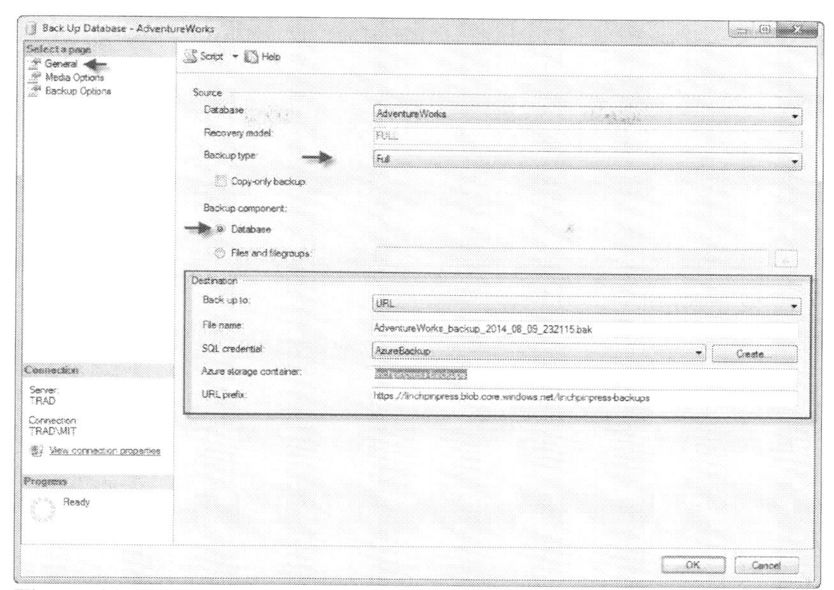

Figure 11.12 Backing up to URL

Using T-SQL

Using T-SQL to back up a database to Azure is similar to using T-SQL to perform any other type of SQL Server backup.

```
BACKUP DATABASE DB_NAME TO <backup_device> WITH <options>
```

The difference is specifying TO URL instead of TO DISK and specifying WITH CREDENTIAL. To back up AdventureWorks to your Azure storage, modify the following script to produce your values.

```
BACKUP DATABASE [AdventureWorks]
TO  URL =
'https://linchpinpress.blob.core.windows.net/linchpinpress
-backups/AdventureWorks_backup_azure_tsql.bak'
WITH CREDENTIAL = N'AzureBackup', COMPRESSION, STATS = 10
```

Because STAT=0, you will receive status updates every 10 percent (as shown in Figure 11.13).

```
Messages
  10 percent processed.
  20 percent processed.
  30 percent processed.
  40 percent processed.
  50 percent processed.
  60 percent processed.
  70 percent processed.
  80 percent processed.
  90 percent processed.
  Processed 24272 pages for database 'AdventureWorks', file 'AdventureWorks2012_Data' on file 1.
  100 percent processed.
  Processed 2 pages for database 'AdventureWorks', file 'AdventureWorks2012_Log' on file 1.
  BACKUP DATABASE successfully processed 24274 pages in 228.090 seconds (0.831 MB/sec).
100 %
```

Figure 11.13 Successful backup of AdventureWorks to Azure storage, using T-SQL

If you browse into your Azure storage account and look in your container, if you performed the UI and T-SQL backups, you will be able to see your two backup files (as in Figure 11.14).

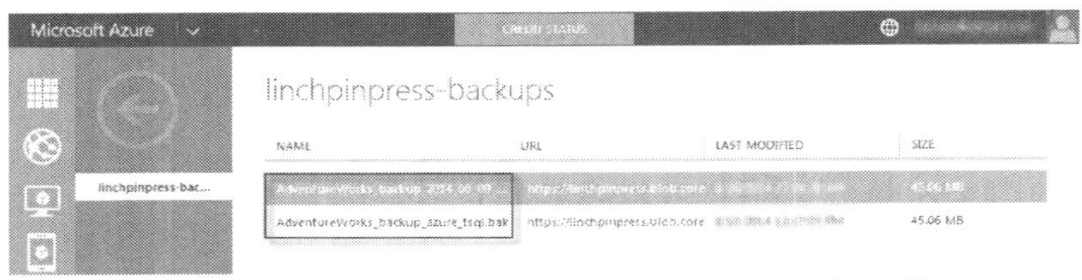

Figure 11.14 Azure storage container showing two backup files.

Restoring a Database from Azure Storage

Restoring a database from Azure storage is similar to restoring a database from disk. The difference is that you specify from URL and provide your Azure information. Begin your restore by right clicking on your database. Choose Tasks > Restore > Database (as in Figure 11.15).

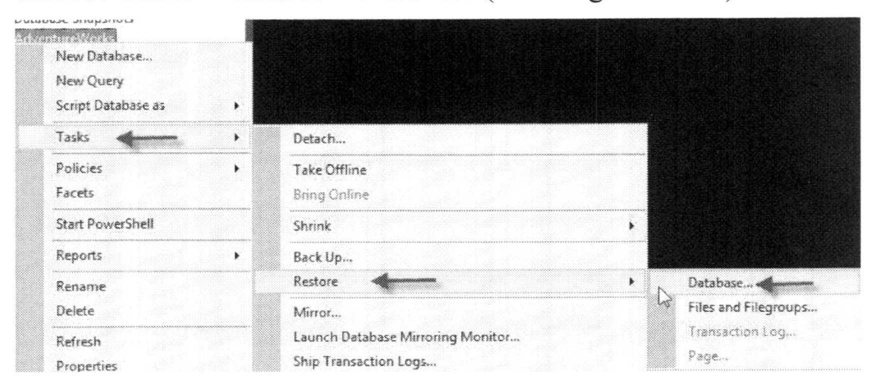

Figure 11.15 Right click on the database. Chose Tasks > Restore > Database

If you are restoring a copy of your backup from Azure on the instance of SQL Server that the backup was made on, you will be prompted to enter your credentials to connect to Microsoft Azure Storage as you see in Figure 11.16.

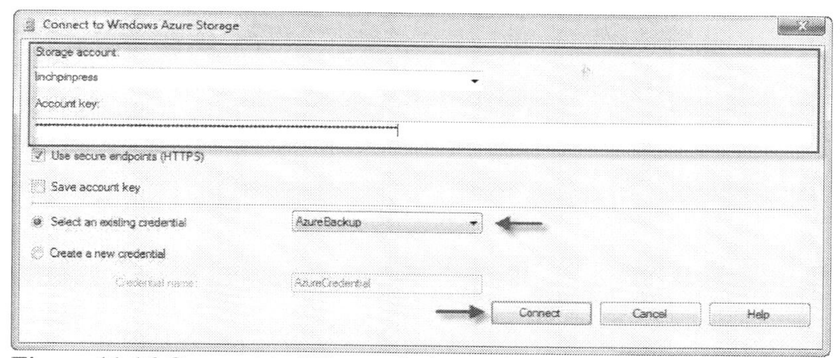

Figure 11.16 Connect to Microsoft Azure Storage

If you're restoring on the server that created the backup, then it will default to wanting to restore the most recent backup, as well as wanting to create a tail log backup prior to performing a restore operation.

If you have been following along with the other demos in this book, you will be familiar with the next set of options. First, click on **Device** as your **Source,** and then click the ellipsis (as shown in Figure 11.17).

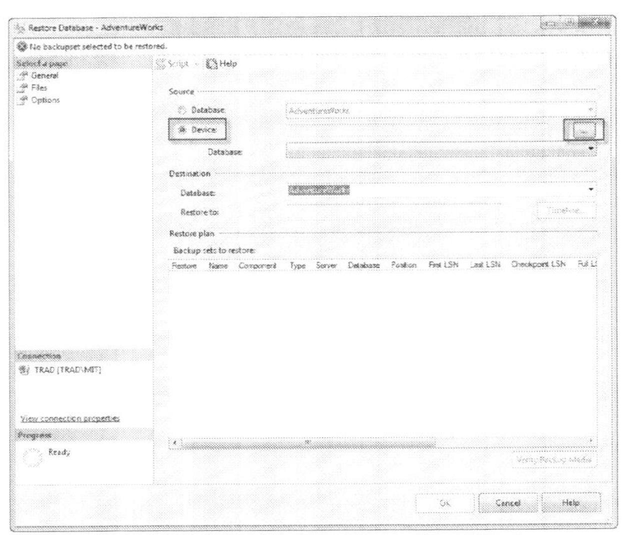

Figure 11.17 Click Device and then the ellipsis.

When the **Select backup devices** box appears, change the **Backup media type** from File to URL, and then click **Add** (as shown in Figure 11.18).

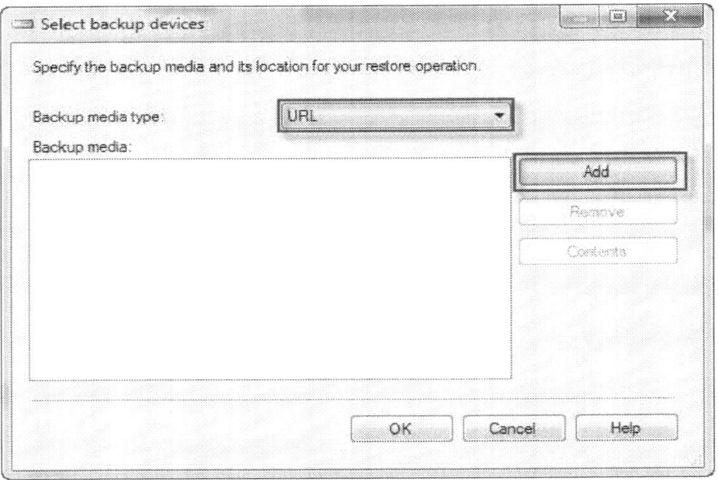

Figure 11.18 Change backup media type to URL, and click Add

The **Locate Backup File in Microsoft Azure** dialog box will appear. From here, you need to expand **Containers** and then click on your container object. This will list the backups so you can choose which one you want to restore. In this example, choose AdventureWorks_backup_azure_tsql.bak (as shown in Figure 11.19), and then click **OK** and **OK** to return to the restore screen.

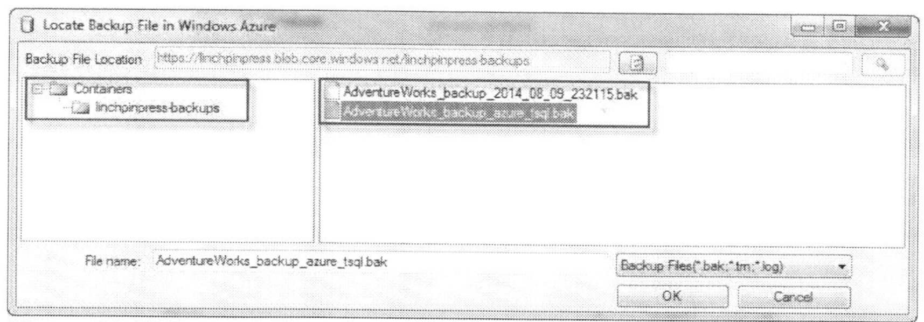

Figure 11.19 Expand Containers, select your container, and pick your database file

Click on **Options**, and uncheck the box for **Take a tail-log backup before restore**, as well as **Overwrite the existing database (WITH REPLACE)** (as shown in Figure 11.20).

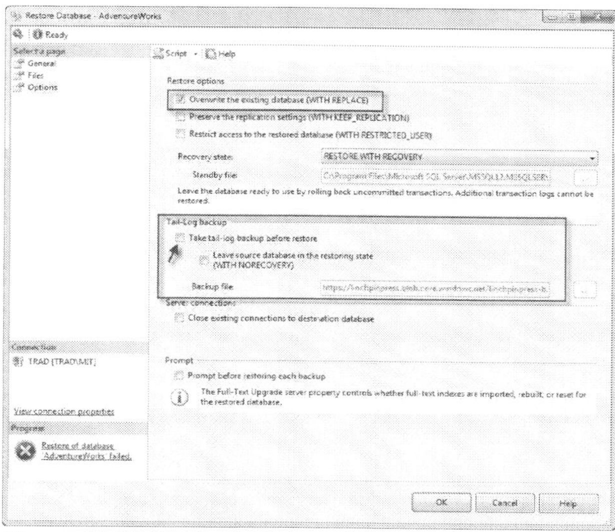

Figure 11.20 Uncheck Take tail-log backup before restore.

Click on the **General** page, and verify your restore settings. Click **OK** to begin the restore. You will see a progression bar in the top right corner (as show in Figure 11.21).

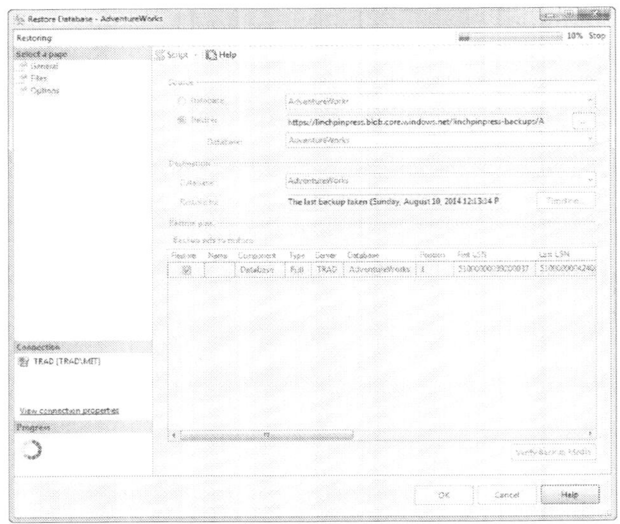

Figure 11.21 Displays database restoring from Azure Storage

Congratulations! You have successfully restored the AdventureWorks database from Azure Storage (as shown in Figure 11.22).

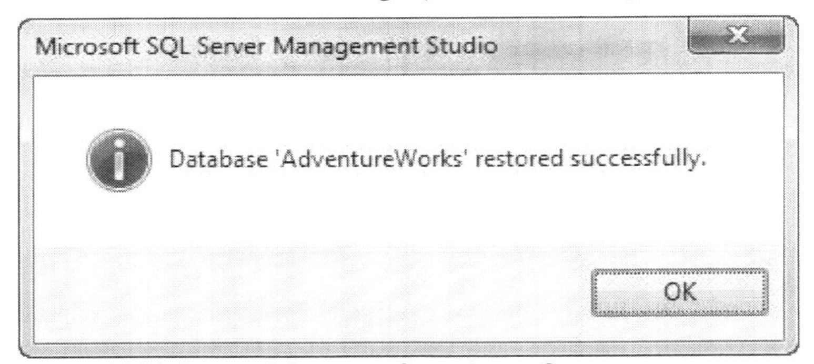

Figure 11.22 Successful restore from Azure Storage

Using T-SQL

Using T-SQL to restore from Azure Storage is similar to performing any other restore using T-SQL. The difference is that you have to specify FROM URL and provide the SQL Server credential name. The restore syntax follows the same standard as any other T-SQL restore, such as:

```
BACKUP DATABASE DB_NAME TO <backup_device> WITH <options>
```

To restore AdventureWorks from your Azure Storage, modify the following script with your values. I have included STATS = 10 to display the restore progress in increments of 10 percent. I also chose to overwrite the existing database.

```
RESTORE DATABASE [AdventureWorks]
FROM  URL =
'https://linchpinpress.blob.core.windows.net/linchpinpress
-backups/AdventureWorks_backup_azure_tsql.bak'
WITH CREDENTIAL = N'AzureBackup', STATS = 10, REPLACE
```

Congratulations! You have now used T-SQL to restore a database from Azure Storage.

Summary

Microsoft Azure Storage is a very cost-effective method for storing your data assets offsite. Microsoft Azure Storage can provide georeplication for redundancy and protection against hardware failures. Georeplicated data means your data is stored in two locations. You chose your primary location when you create your account and the secondary location is automatically determined based of the location of your primary. In addition, Microsoft Azure Storage can provide a cost-effective offsite and backup archive solution.

To back up your SQL Server databases to Azure Storage, you need an Azure account. Create a Storage account and a storage container.

To configure SQL Server to back up to Azure Storage, create a SQL Server Credential using your Azure Storage account information. Once you have established the credential, you can back up to URL and specify **WITH CREDENTIAL**.

Points to Ponder for Backup to Azure Storage

1. Microsoft Azure Storage provides a cost-effective way to have your backups georeplicated.
2. The speed of backing up and restoring to Microsoft Azure Storage depends on your Internet connection.
3. Microsoft Azure Storage can help satisfy audit or compliance requirements for your organization by providing offsite storage.

Review Quiz – Chapter Eleven

1. To back up SQL Server databases to Azure Storage, you must first create the credentials in SQL Server to connect to Azure.
    a. TRUE
    b. FALSE
2. To create a Microsoft Azure Storage account to back up SQL Server database, you need which of the following? Choose all that apply.
    a. Microsoft Azure account
    b. Storage account
    c. Container
    d. Virtual machine
3. Microsoft Azure Storage can provide georeplicated data.
    a. TRUE
    b. FALSE

Answer Key

1. For SQL Server to connect to Microsoft Azure, you must define the credential within SQL Server; therefore A is the answer.
2. To create a Microsoft Azure Storage account, you will need a Microsoft Azure account, storage account, and container; therefore A, B, and C are correct.

3. Microsoft Azure Storage can provide georeplicated data; therefore A is the correct answer.

INDEX

R

S

T

V

W

Made in the USA
Lexington, KY
08 October 2014